NITRIFICATION AND DENITRIFICATION

Edited by **Ivan X. Zhu**

Nitrification and Denitrification
http://dx.doi.org/10.5772/65170
Edited by Ivan X. Zhu

Contributors

Yiu Fai Tsang, Marlene Roeckel, Víctor Guzmán-Fierro, Constanza Arriagada, Makoto Shoda, Marjana Simonic, Andreja Goršek, Aleksandra Petrovič, Emir Martínez Gutiérrez, Anne-Claire Texier, Flor De María Cuervo Lopez, Jorge Gómez, Ivan X Zhu

Published by InTech

Janeza Trdine 9, 51000 Rijeka, Croatia

Publishing Process Manager Marijana Francetic
Technical Editor SPi Global
Cover InTech Design team

First published June, 2017
Printed in Croatia
Legal deposit, Croatia: National and University Library in Zagreb

Additional hard copies can be obtained from orders@intechopen.com

Nitrification and Denitrification, Edited by Ivan X. Zhu
p. cm.
Print ISBN 978-953-51-3299-8
Online ISBN 978-953-51-3300-1

For EU product safety concerns:
IN TECH d.o.o., Prolaz Marije Krucifikse Kozulić 3, 51000 Rijeka, Croatia, info@intechopen.com or visit our website at intechopen.com.

Meet the editor

Dr. Ivan Zhu has highly specialized expertise in biological wastewater treatment, membrane applications to industrial and municipal water and wastewater treatment, flocs and biofilm characterization in terms of microbial community distribution and extracellular polymeric substances, and membrane fouling characterization. He has applied his extensive knowledge of separation processes to evaluation and design of water and wastewater chemical/biological treatment processes. Previously, he worked at Xylem Water Solutions, where he gained extensive experience in drinking water treatment, wastewater tertiary treatment, denitrification, biological active filtration, ozone-enhanced biofiltration, and dissolved air flotation. Presently, he is working at Evoqua Water Technologies as an application engineer for integrated industrial solutions for water and wastewater treatment. He holds a bachelor's degree from Shanghai Jiao Tong University in Shanghai, China, and master's and doctoral degrees from the University of Toronto, Ontario, Canada.

Contents

Preface

With the application of nitrate-containing fertilizers, consumption of animal products, and industrial production activities, ever more ammonia and nitrate are being discharged into rivers and lakes, which may cause eutrophication and deterioration of aquatic environments. The maximum acceptable contamination level in drinking water is 10 mg/L nitrate nitrogen in the USA, Japan, and Korea, while the European Union countries set the standard for nitrate nitrogen at 11.3 mg/L, and the World Health Organization recommends 11.3 mg/L nitrate nitrogen to protect against methemoglobinemia in bottle-fed infants. To protect aquatic ecological systems, an ever more stringent limit was imposed to point source dischargers into sensitive water bodies, such as Chesapeake Bay area in the USA. Nitrification and denitrification are the fundamental process routes in nitrogen removal in aquatic ecosystems. They play an essential role in natural systems and engineered systems in terms of the nitrogen cycle. This book provides a state-of-the-art overview and discusses the fundamental research on nitrogen removal. Some case studies of full-scale applications are also covered.

Topics will broadly include, but will not be limited to:

- Biodegradation of ammonia and the analysis of microbial communities
- Nitrification and denitrification in natural aquatic systems
- Simultaneous nitrification and denitrification in lab-scale and full-scale applications
- Nitrogen management strategies of industrial wastewater such as fishery
- Heterotrophic nitrification and aerobic denitrification and the identification of microbial species
- Anaerobic ammonia oxidation (ANAMMOX)
- Advanced technology such as membrane bioreactors for groundwater nitrogen removal

Nitrification and denitrification are fundamental for nutrient removal. While extensive research had been conducted, this book is oriented to innovative processes and selected applications such as heterotrophic nitrification and anaerobic ammonia oxidation. It is anticipated that this book shed light for future research and innovation. This book is extremely useful for design engineers, researchers, and practitioners.

Finally, during the course of editing and compiling this book, extensive support and guidance were received from Ms. Marijana Francetic, publishing process manager. The editor would like to express deep appreciation and gratefulness for her support.

Dr. Ivan X. Zhu
University of Toronto
Toronto, Canada

Chapter 1

Introductory Chapter: Effects of Salinity on Biological Nitrate Removal from Industrial Wastewater

Ivan X. Zhu and Jian R. Liu

Additional information is available at the end of the chapter

http://dx.doi.org/10.5772/intechopen.69438

1. Introduction

Nitrification and denitrification is a major process route for nitrogen cycles in ecological systems. It is essential for nitrogen removal from water and wastewater. In the past, there were tremendous efforts and a significant amount of research on this topic in regard to microbial species, effects of operating parameters such as temperature, pH, dissolved oxygen (DO), types of carbon sources, and hydraulic and mass loadings. In this book, a comprehensive review of conventional technologies was carried out and innovative technologies such as *anaerobic ammonium oxidation* (*Anammox*) were focused upon. A unique species *Alcaligenes faecalis* No. 4 was reviewed and experimented for heterotrophic nitrification and aerobic denitrification. Aerobic biofiltration for ammonia removal and an anoxic membrane bioreactor (MBR) for nitrate removal were selected as application examples.

However, the effect of salinity on nitrification and denitrification was not discussed. Some industrial wastewater contains high concentrations of salinity and nitrate, that is, flue-gas desulfurization (FGD) wastewater contains nitrate (about 50 mg/L as N), high total dissolved solids (TDSs) (1–5% of chloride), and other contaminants. Elevated TDS levels present in industrial wastewater may have profound effect on denitrification and there had been little research on this subject matter. To remove nitrate from such wastewater, it is important to understand the effects of salinity on the process kinetics, selection of carbon sources, periodical salinity fluctuations, and microbial communities for the process selection and engineering design.

2. Selection of carbon sources in process design

Heterotrophic denitrification occurs in the presence of both nitrate and biodegradable organic substances under anoxic conditions. If a denitrification system is placed after a secondary wastewater treatment process, intrinsic biodegradable organic substances are essentially

Carbon source	COD/N	Yield (gVSS/gCOD)	μ_{max} (d^{-1})	k_D (mgN/gVSS-h)
Methanol	4.1–4.5	0.23–0.25	0.77 (15°C) 2 (20°C)	32 (15°C) 91 (20°C)
Ethanol	5.9	0.25–0.28	1.89 (15°C) 4.8 (25°C)	46 (15°C) 139 (20°C)
Acetate	5.7	0.35	1.2 (13°C) 3.5 (19°C)	13.6
Glucose	8.9	0.38		3.8

Table 1. Kinetic information of selected carbon sources [1].

depleted before the denitrification unit. Under these situations, external supplementation of organic substances (electron donors) is usually needed to generate dedicated microbial communities. Generally, an external carbon source, such as methanol, ethanol, acetic acid, glycerol, sugar, or molasses, is used as a supplement. Another commercial product worth mentioning is MicroC™. It is manufactured by Environmental Operating Solutions Inc. (Bourne, MA) and is an environmentally benign, proprietary wastewater treatment chemical containing a mixture of organic compounds, mainly glycerol. It contains 670,000 mg/L chemical oxygen demand (COD) with a specific gravity of 1.22 g/mL at 25°C.

The stoichiometric reaction C/N ratio, yield, specific growth rate, and Arrhenius temperature factor are different for different carbon sources, some of which are summarized in **Table 1**. It should be advised that these parameters were obtained with municipal wastewater under different acclimation and feeding conditions and microbial compositions, and the use of these parameters should be with care.

3. Impact of salinity on specific denitrification rate

Specific denitrification rates (SDNRs) are usually expressed in the mass of nitrate removed within a unit time in regard to one unit of reactor volume, biomass, biofilm surface, or fixed-film media bed. There had been conflicting reports about the effects of salinity on specific denitrification rates.

Osaka et al. [2] studied two suspended biomass systems fed with acetate acid and methanol, respectively, and found that acetate-fed process attained high nitrate removal at 0–10% NaCl, whereas methanol was shown effective for nitrate removal at 0–3% NaCl without sacrificing efficiencies. Nitrate removal efficiencies were close to 100% at a mass loading of 0.15 g NO_3-N/g MLSS/day or a volumetric loading of 0.75 kg NO_3-N/m^3/day. This study was carried out in a manner that allowed enough time (at least 20 days) for microbial communities to adapt to a higher salinity with a 1% incremental change.

Similar to the observation by Osaka et al. [2], the denitrification rate with methanol as a carbon source was unaffected by sodium chloride up to 2% in a fluidized bed biofilm reactor with media carriers encapsulated with mixed denitrification cultures [3].

SDNR was 0.06 g NO_3-N/g MLSS/day for a freshwater system without salt spiking; SDNR appeared not to be affected (similar to 0.06 g NO_3-N/g MLSS/day) for a system with 5 g/L salt spiking, and it only slightly decreased to 0.048 g NO_3-N/g MLSS/day for an acclimated system with 30 g/L salt addition [4]. In fully acclimated systems (two bench-scale sequencing batch reactors operated in parallel for 4 months), as complete denitrification occurred, the maximum specific nitrate reduction rate was 1.2 g $NO_3 \pm$ N/g MLSS/day at a wastewater TDS concentration of 4.8% with acetate as a carbon source and the denitrification rate was decreased to 0.456 g $NO_3 \pm$ N/g MLSS/day at 18% TDS [5]. These studies suggest that acclimated (to saline water) systems appeared less sensitive to salinity increase.

The maximum nitrification and denitrification rates were 0.05 and 0.036 g NO_3-N/g VSS/day, respectively, in a down-flow hanging sponge reactor treating phenol (electron donor for denitrification) and ammonia wastewater. The system had been acclimated for 1100 days with 10.9 g/L chloride before the study where a dominant species, *Azoarcus*-like species, was found [6]. The maximal denitrification rate achieved with ethanol mixture (industrial byproduct) (0.64 g N-NO_x/g VSS/d) was much higher than the rate reached with methanol mixture (industrial byproduct) (0.11 g N-NO_x/g VSS/d) at sulfate concentrations of 1.5–2% after 450 days of operation [7]. Pure culture *Pseudomonas stutzeri* in a packed bed bioreactor achieved high denitrification rate of 0.84 kg NO_3-N/(m^3/day) or 0.025–0.13 g NO_3-N/g biomass/day at 10 g/L salinity [8]. The strain PAD-2 (closely related to *M. alkaliphilus*) in genus *Marinobacter* of *γ-proteobacteria* exhibited higher denitrification rates at concentrations of 3–6% than at other salinities of 12–18% w/w [9].

On the contrary to the above studies, it was concluded that denitrification rates were severely affected with salt spiking. At 1.52% of salt spiking, a specific denitrification rate decreased by half from 0.7 to 0.35 kg NO_3-N/m^3/day [10]. In another study by Ucisik and Henze [11], it was found that a specific denitrification rate decreased with an increasing chloride concentration in a suspended growth system fed with acetate, and the maximal specific denitrification rate decreased from 1.2 kg NO_3-N/m^3/day at 0.48% chloride down to 0.04 kg NO_3-N/m^3/day at 9.67% chloride. However, this study may still have suffered from insufficient acclimation time, as at each chloride concentration level, the microorganisms were only allowed to acclimatize for 4–5 days. The spiking of salt sharply reduced the microbial activity in an activated sludge system seeded with municipal sludge. When salt concentrations were below 10 g/L NaCl, microorganisms were able to acclimatize in several weeks and achieve the same initial activity as in raw sludge samples; when the salt concentration was above 30 g/L NaCl, the acclimatization process was slow [12]. A mathematical model was developed to predict the SDNR at different salt spiking levels where a salt inhibition constant was identified to be 1.52% (SDNR was reduced by half) [10].

Table 2 summarizes SDNR in high-salinity wastewater and SDNR varied from 0.75 to 4.8 kg NO_3-N/m^3/day or 0.025 to 1.2 g/g biomass/day, depending on the salinity levels, carbon sources, and temperature. It appeared that biofilm systems had relatively higher volumetric denitrification rates as compared to the suspended growth systems. A maximal denitrification rate of 4.8 kg NO_x-N/m^3 media bed/day(sintered fly ash) was achieved in a fluidized bed reactor; 2.5 kg NO_3-N/m^3/day was achieved with a reactor filled with sponge cubes for microbial attachment; and 0.84 kg NO_3-N/m^3/day was achieved in a packed bed reactor (with clinoptilolite). These observations of high rates were perhaps attributed to higher specific surface area of carrier media and higher biomass density. Furthermore, in a biofilm reactor filled with cellulose triacetate carriers encapsulated with mixed denitrification cultures, an exceptionally high denitrification rate of 11 kg/m^3 media bed/day was achieved [3].

Denitrification rate	Acclimation and culture	Carbon source	Salinity	System	Reference
0.84 kg NO_3-N/(m^3/day) at 10 g/L salinity; 0.025–0.13 g NO_3-N/(g biomass.day)	*Pseudomonas stutzeri*	Ethanol	10–40 g/L	Packaged bed system (clinoptilolite)	[8]
0.75 kg NO_3-N/m^3/day or 0.15 g NO_3-N/g MLSS/day (10% salinity with acetic acid)	The saline concentration was steadily increased by 1% salinity with NaCl from 0%; at each salinity level, at least 20 days were maintained	Acetate and methanol	0–100 g/L	Suspended growth system	[2]
1.2 kg NO_3-N/m^3/day at 4.8 g/L chloride; 0.04 kg NO_3-N/m^3/day at 96.7 mg/L chloride	At each chloride level, 4–5 days were allowed for acclimation	Acetate	4.8–96.7 g/L	Suspended growth system	[11]
0.7 kg NO_3-N/m^3/day for 0% NaCl and 0.35 for 1.52% NaCl	Spiking	Sugar	0–6%	Packaged bed system (1 cm plastic tubes)	[10]
A slight drop in nitrogen removal, NR, and DNR was observed, when the salinity was increased from 4.2 to 9.8 g NaCl/L		Intrinsic COD	4.2–9.8 g NaCl/L	Sequential batch biofilm reactor	[17]
2.5 kg NO_3-N/m^3/day at 10% salinity	*Halomonas* sp. and *Marinobacter* sp.; seed sludge was acclimated for 3 years	Acetate	2 and 10%	Sponge cubic media	[14]
0.8 kg NO_3-N/m^3/day	*P. pantotrophus and P. fluorescens*	Biodegradable hydrocarbons Brenntaplus VP1	Up to 35 g/L Cl^- and 17 g/L SO_4^{2-}	Bacteria encapsulated in porous polyvinyl alcohol lenses	[15]
0.036 g/g-VSS/day	*Azoarcus*-like species; acclimated for 1100 days prior to the study	Phenol	10.9 g Cl^-/L	Down-flow hanging sponge reactor	[6]
0.64 g N-NO_x/g VSS/d with ethanol; 0.11 g N-NO_x/g VSS/d with industrial waste methanol	The two-sludge plant was operated continuously for 450 days, using real, high-strength industrial wastewater	Industrial ethanol mix; industrial methanol mix	1.5–2.0% SO_4^{2-}	Suspended growth system	[7]
1.2 g NO_3 ± N/g MLSS/day at TDS 4.8%; 0.456 g NO_3 ± N/g MLSS/day at 18% TDS	Reactors operated in parallel for 4 months	Acetate	4.8, 16, and 18%	Suspended growth system	[5]
4.8 kg NO_x-N/m^3 media bed/day	Acclimated	Acetic acid	45 g/L Cl^-	Fluidized bed system	[18]
11 kg NO_x-N/m^3 media bed/day or 4.8 kg NO_x-N/m^3/day	Media carriers encapsulated with mixed culture	Methanol	0–30 g/L NaCl	Fluidized bed system	[3]

Denitrification rate	Acclimation and culture	Carbon source	Salinity	System	Reference
0.06 g NO_3-N/g MLSS/day without salt addition; 0.06 g NO_3-N/g MLSS/day for 5 g/L salt and 0.048 g NO_3-N/g MLSS/day for 30 g/L salt	Acclimated	Sucrose and acetic acid	30 g/L NaCl	Suspended growth system	[4]
0.305 (on acetate); 0.36 (on lactate), 0.39 (on glycerol), and 0.045 (on ethanol) g NO_3-N/g biomass/day	*Halomonas campisalis* sp. *Nov.*		12.5% NaCl	Suspended growth system	[16]

VSS, volatile suspended solids; MLSS, mixed liquor suspended solids; TDS, total dissolved solids; COD, chemical oxygen demand.

Table 2. Denitrification rates under different conditions.

In summary, it appeared that denitrification efficiency will drop upon an initial increase of salinity and can be sustained if biomass is properly acclimated and adapted to corresponding salinities, and rates were comparable to that at low-salinity concentrations. However, in full-scale installations, this effect may be pronounced during the initial period of commissioning, which usually required the designer to provide enough redundancy for the process, or seed the process with an acclimated culture obtained elsewhere to speed up the process.

4. Halophilic cultures

Halophilic bacteria are microorganisms that do not need sodium chloride to grow but can grow in high-salinity environments. Halophilic bacteria are classified into three groups according to their response to sodium chloride concentrations: (i) the slight halophiles (most rapid growth at 2–5% NaCl), (ii) the moderate halophiles (most rapid growth at 5–20% NaCl), and (iii) the extreme halophiles (most rapid growth at 20–30% NaCl) [13].

The phylogenetic analysis showed that the strains isolated from acclimated sludge (to saline water) had a high similarity to the genus *Alcaligenes* in *β-proteobacteria* and the genera *Vibrio, Pseudomonas, and Halomonas* in *γ-proteobacteria*. Genera *Halomonas and Marinobacter* in *γ-proteobacteria* were isolated [14]. *α-Proteobacteria* were also found [6]. *Azoarcus*-like species in *β-proteobacteria* was identified to conduct denitrification using phenol [6]. It was found that the dominant species shifted when salinity varied [14].

Researchers used microorganisms *P. fluorescens* and *P. pantotrophus* for denitrification in saline water [15]. *P. stutzeri* in the packed bed bioreactor achieved a high denitrification rate at 10 g/L salinity [8], and the strain PAD-2 (closely related to *M. alkaliphilus*) in genus *Marinobacter* of *γ-proteobacteria* also exhibited high denitrification rates at concentrations of 3–6% [9]. The species

M. aquaeolei sp. *nov.* was found to grow under anoxic conditions in the presence of nitrate on succinate, citrate, or acetate, but not on glucose. It was also interesting that *H. campisalis* sp. *nov.* grew on acetate, lactate, glycerol, and ethanol but not on methanol [16].

Table 3 summarizes the species capable of denitrifying under saline conditions. Most of the species were in the class of *γ-proteobacteria*. Species were found to even survive in a wide range of salinity as high as 23.4%.

5. Summary and future perspective

Extensive research was conducted in the past for denitrifiers in treating municipal wastewater which typically contains TDS less than 1000 mg/L. Whether the knowledge acquired in regard to the stoichiometry, kinetics, metabolic pathways, and microbial communities and characteristics is transferrable to halophilic counterparts is subject to further research.

Through the above short review, it was suggested that specific denitrification rates of mixed cultures decreased with an increasing salinity concentration. However, some specific species such as *H. campisalis* sp. *nov.* exhibited relatively high denitrification rates at 12.5% salinity with different carbon sources, similar to that of its freshwater counterparts. If biomass is properly acclimated and adapted to saline environments, the SDNR could be comparable to that at low-salinity concentrations.

Class	Genus	Species	Salinity range	Optimal salinity	Reference
γ-Proteobacteria	*Halomonas*	*campisalis* sp. *Nov.*	1.17–26.3 % (w/v)	8.8% (w/v)	[16]
	Halomonas	*daqingensis* sp. *nov.*	1.0–15.0% (w/v)	5–10% (w/v)	[19]
	Halomonas	*ventosae* sp. *nov.*	1.0–15.0 % (w/v)	8% (w/v)	[20]
	Halomonas	*chromatireducens* sp. *nov.*	0.585–23.4% (w/v)	2.9% (w/v)	[21]
	Halomonas	*desiderata* sp. *nov.*	0–18% (w/w)		[22]
	Marinobacter	*aquaeolei sp. nov.*	0–20% (w/w)	5% (w/w)	[23]
	Marinobacter	*PAD-2 (closely related to M. alkaliphilus)*	3–18% (w/w)	3–6% (w/v)	[9]
	Pseudomonas	*stutzeri*	1–4% (w/v)		[8]
	Pseudomonas	*pantotrophus and fluorescens*	Up to 3.5% (w/v) Cl^- and 1.7% (w/v) SO_4^{2-}		[15]
β-Proteobacteria	*Azoarcus*	*Azoarcus*-like species	1.09% (w/v) Cl^-		[6]

Table 3. Summary of halophilic denitrifying species.

Author details

Ivan X. Zhu[1*] and Jian R. Liu[2]

*Address all correspondence to: ivan.zhu@evoqua.com

1 Evoqua Water Technologies, Warrendale, PA, USA

2 WSP, Markham, ON, Canada

References

[1] Omnis-Hayden A, Gu AZ. Comparison of organic sources for denitrification: Biodegradability, denitrification rates, kinetic constants and practical implication for their application in WWTPs, In: WEFTEC, Chicago; October 20-22, 2008; WEF: Chicago; 2008. pp. 253-273

[2] Osaka T, Shirotani K, Yoshie S, Tsuneda S. Effects of carbon source on denitrification efficiency and microbial community structure in a saline wastewater treatment process. Water Research. 2008;**42**(14):3709-3718

[3] Yang PY, Nitisoravut S, Wu JS. Nitrate removal using a mixed culture entrapped microbial cell immobilization process under high salt conditions. Water Research. 1995;**29**(6):1525-1532

[4] Panswad T, Anan C. Impact of high chloride wastewater on an anaerobic/anoxic/aerobic process with and without inoculation of chloride acclimated seeds. Water Research. 1999;**33**(5):1165-1172

[5] Glass C, Silverstein J. Denitrification of high nitrate, high salinity wastewater. Water Research. 1999;**33**(1):223-229

[6] Ikeda N, Natori T, Okubo T, Sugo A, Aoki M, Kimura M, Yamaguchi T, Harada H, Ohashi A, Uemura S. Enhancement of denitrification in a down-flow hanging sponge reactor by effluent recirculation. Water Science and Technology. 2013;**68**(3):591-598

[7] Carrera J, Baeza JA, Vicent T, Lafuente J. Biological nitrogen removal of high-strength ammonium industrial wastewater with two-sludge system. Water Research. 2003;37(17): 4211-4221

[8] Rezaee A, Godini H, Soltani RDC, Hossini H, Nili-Ahmadabad A. Denitrification of high salinity, high nitrate wastewater using clinoptilolite in a packed bed bioreactor. Journal of Environmental Research and Management. 2012;**3**(2):031-036

[9] Nakano M, Inagaki T, Okunishi S, Tanaka R, Maeda H. Effect of salinity on denitrification under limited single carbon source by *Marinobacter* sp. isolated from marine sediment. Journal of Basic Microbiology. 2010;**50**(3):285-289

[10] Dincer AR, Kargi F. Salt inhibition of nitrification and denitrification in saline wastewater. Environmental Technology. 1999;**20**(11):1147-1153

[11] Ucisik AS, Henze M. Biological denitrification of fertiliser wastewater at high chloride concentration. Water South Africa. 2004;**30**(2):191-195

[12] Linaric M, Markic M, Sipos L. High salinity wastewater treatment. Water Science and Technology. 2013;**68**(6):1400-1405

[13] Ollivier B, Caumette P, Garcia JL, Mah RA. Anaerobic bacteria from hypersaline environments. Microbiology Review. 1994;**58**(1):27-38

[14] Yoshie S, Ogawa T, Makino H, Hirosawa H, Tsuneda S, Hirata A. Characteristics of bacteria showing high denitrification activity in saline wastewater. Letters in Applied Microbiology. 2006;**42**(3):277-283

[15] Trogl J, Krhutkova O, Pilarova V, Danova P, Holıcek R, Kohlova M, Hejda S, Smrcka J, Bouskova A, Kriklavova L. Removal of nitrates from high-salinity wastewaters from desulphurization process with denitrifying bacteria encapsulated in Lentikats biocatalyst. International Journal of Environmental Science & Technology. 2012;**9**(3):425-432

[16] Mormile MR, Romine MF, Garcia MT, Ventosa A, Bailey TJ, Peyton BM. *Halomonas campisalis* sp. nov., a denitrifying, moderately haloalkaliphilic bacterium. Systematic and Applied Microbiology. 1999;**22**(4):551-558

[17] She ZL, Zhang XL, Gao MC, Guo YC, Zhao LT, Zhao YG. Effect of salinity on nitrogen removal by simultaneous nitrification and denitrification in a sequencing batch biofilm reactor. Desalination and Water Treatment. 2016;**57**(16):7378-7386

[18] Vredenbregt LHJ, Nielsen K, Potma AA, Kristensen GH, Sund C. Fluid bed biological nitrification and denitrification in high salinity wastewater. Water Science and Technology. 1997;**36**(1):93-100

[19] Wu G, Wu XQ, Wang YN, Chi CQ, Tang YQ, Kida K, Wu XL, Luan ZK. *Halomonas daqingensis* sp. nov., a moderately halophilic bacterium isolated from an oilfield soil. International Journal of Systematic and Evolutionary Microbiology. 2008;**58**(12):2859-2865

[20] Martínez-Cánovas MJ, Quesada E, Llamas I, Béjar V. *Halomonas ventosae* sp. nov., a moderately halophilic, denitrifying, exopolysaccharide producing bacterium. International Journal of Systematic and Evolutionary Microbiology. 2004;**54**(3):733-737

[21] Shapovalova AA, Khizhniak TV, Turova TP, Sorokin DY. *Halomonas chromatireducens* sp. nov., a new denitrifying facultatively haloalkaliphilic bacterium from solonchak soil capable of aerobic chromate reduction. Microbiology. 2009;**78**(1):102-111

[22] Berendes F, Gottschalk G, Heine-Dobbernack E, Moore ERB, Tindall BJ. *Halomonas desiderata* sp. nov., a new alkaliphilic, halotolerant and denitrifying bacterium isolated from a municipal sewage works. Systematic and Applied Microbiology. 1996;**19**(2):158-167

[23] Huu NB, Denner EB, Ha DT, Wanner G, Stan-Lotter H. *Marinobacter aquaeolei* sp. nov., a halophilic bacterium isolated from a Vietnamese oil producing well. International Journal of Systematic Bacteriology. 1999;**49**(2):367-375

Chapter 2

Innovative Nitrogen and Carbon Removal

Marlene Roeckel von Bennewitz,
Constanza Beatriz Arriagada Gajardo and
Víctor Gabriel Guzmán-Fierro

Additional information is available at the end of the chapter

http://dx.doi.org/10.5772/68054

Abstract

The aerobic systems have been the most widely biological treatment used for municipal and industrial wastewater but show serious problem with sludge sedimentation, high energy demand and microbial inhibition. On the other hand, the anaerobic digestion (AD) of wastewater is considered the best alternative to remove the organic compounds and to recover energy via methane production. Nevertheless, AD has a problem: the treatment of industrial wastewater with high organic nitrogen content reaches high free ammonia (NH_3) concentrations due to the protein degradation. NH_3 inhibits the methanogenic process and is toxic to the environment, and then, it must be removed before its final disposition. Several physicochemical processes have been evaluated for the recovery or/and treatment of ammonium from wastewater. The most frequent treatments are gas stripping and magnesium ammonium phosphate precipitation. These methods are effective, but they are very expensive compared to biological treatments. Moreover, these techniques usually require more power consumption than the biological process. The technologies based on partial nitrification and Anammox (PN-A) are the ones with better performance. Thus, this chapter mainly focuses on biological processes based on AD, denitrification and PN-A for the removal of carbon and nitrogen from industrial wastewater with recovery of energy and water.

Keywords: anammox, anaerobic digestion, nitrogen removal, carbon removal, partial nitrification, REMON

1. Introduction

Anaerobic digestion (AD) of high load wastewater is considered one of the best alternatives to remove the organic compounds and to recover energy via the production of methane, which has significant advantages over other forms of bioenergy production. The bottleneck of many industry wastewater AD is the high content of generated total ammonia nitrogen (TAN, ammonium + ammonia) reaching inhibitory concentrations for methanogenic bacteria, which result in suboptimal production of methane. Anaerobic reactors fed with high ammonia concentrations also produce an effluent with high TAN concentrations, which require to be treated [1].

In addition, physicochemical processes have been evaluated for the recovery and/or treatment of ammonia from wastewaters. Recovery is usually done with struvite precipitation; on the other hand, the most common treatment is gas stripping [2]. Nevertheless, these processes require the addition of chemicals and a previous carbonate treatment to avoid the precipitation on the equipment. Therefore, the physicochemical treatment is more expensive than the biological treatment [3].

Among several biological processes for the abatement of nitrogen species, we will discuss the different biological technologies based on AD, denitrification, partial nitrification and anaerobic ammonium oxidation (Anammox). Most methods can be applied to treat municipal wastewater, agricultural residues and high nitrogen wastewaters from chemical processes.

The classic biological treatment for nitrogen removal from wastewaters has been the coupled nitrification/denitrification processes, but, in the last 20 years, the partial nitrification-Anammox (PN-A) technology has proved to be efficient in nitrogen removal [3]. The PN-A process is a completely autotrophic technology that compared with the conventional nitrification/denitrification process shows many advantages: (1) consumes 60% less oxygen since a partial nitritation is needed; (2) produces 85% less of sludge mainly due to the slow biomass growth of autotrophic bacteria; (3) no organic matter is needed, which makes it an excellent process to use with anaerobic digestion and (4) releases less greenhouse gases (CO_2, N_2O, etc.) to the atmosphere [4–8]. Even more, in this chapter, we show a new biological technology using the concept of AD and PN-A with water reuse that reduces cost from the annual fresh water consumption and heating.

2. Ammonium rich wastewaters

Industrial activities, summarized in **Table 1**, generate wastewaters rich in organic matter expressed as chemical oxygen demand (COD) and/or a nitrogen-rich wastewater (expressed as TAN, total Kjeldhal nitrogen (TKN) or total nitrogen (TN)) [9–29]. High levels of TAN discharged to the environment can cause serious damage. Emissions of TAN on aquatic systems cause hypoxia: ammonia is oxidized to nitrite and nitrate, promoting biomass growth (mainly algal biomass), and then, eutrophication of water bodies occurs, affecting aquatic life and decreasing the biodiversity because of low availability of dissolved oxygen [30]. Moreover, nitrogen contamination can affect even human health. Consumption of polluted water with nitrate can lead reproductive diseases, methaemoglobinaemia and cancer [31]. Thus, environmental regulations set maximum values allowed to release into the environment.

AD process is an excellent alternative to treat wastewaters with high COD concentration (>3000 mg COD/L) because it does not require oxygen and has low sludge production, and

Raw effluents				
Effluent	**COD (mg/L)**	**TAN (mg/L)**	**TKN (TN) (mg/L)**	**Reference**
Fish industry effluent	5000–32,000	39–1940	n.d.	[9]
Winery wastewater	50	4000–6000	n.d.	[10]
Olive mill wastewater	40,300 ± 1000	n.d.	(240 ± 50)	[11]
Optoelectronic industrial wastewater	13.5 ± 0.7	3712 ± 120	3799 ± 9	[12]
Swine wastewater	3000–15,000	400–1400	(600–2100)	[13]
Cheese whey	73,000–86,000	58–150	(897–1200)	[14]
Tannery wastewater	2400–2600	200–230	n.d.	[15]
Abattoir wastewater	5800–6100	130–280	530–810	[16]
Domestic sewage WWTP	160–320	47–76	(50–89)	[17]
Coke wastewater	630–6500	50–400	250–550	[18]
Poultry manure	43,000 ± 4800	2443 ± 260	n.d.	Our group, unpublished
Piggery wastewater	19,990 ± 2458	740 ± 56	n.d.	[19]
Pharmaceutical wastewater	415–843	123–257	n.d.	[20]
Recycled fish meal effluents	5000–6300	n.d.	480–800	[21]
Brewery wastewater	1300–2300	15–28	(30–37)	[22]
Glass	n.d.	300–650	n.d.	[29]
Coal gasification	n.d.	<1000	n.d.	[29]
Explosives	n.d.	<1503	n.d.	[29]
Landfill leachate	554 ± 97	634 ± 143	n.d.	[23]
Monosodium glutamate wastewater	25,000 ± 5000	19,000 ± 1000	n.d.	[24]
Anaerobic digestion effluents (ADE)				
Effluent	**DQO (mg/L)**	**TAN (mg/L)**	**TKN (TN) (mg/L)**	**Reference**
ADE of sludge	1500–2000	800–900	n.d.	[25]
ADE of fish canning wastewater	914 ± 291	324 ± 36	n.d.	[26]
ADE of abattoir wastewater	800 ± 200	1388 ± 70	n.d.	[16]
ADE of poultry manure	11,860 ± 1270	2533 ± 326	n.d.	Our group, unpublished data
ADE of piggery farm	1980	1200	(1240)	[27]
ADE of slaughterhouse	544–3240	485–783	n.d.	[28]

TAN, total ammonia nitrogen; TKN, total kjeldhal nitrogen; TN, total nitrogen; n.d., not determined.

Table 1. Organic matter and nitrogen rich industry wastewaters.

it is a sustainable process because of the biogas production. Nevertheless, the TAN concentration increases during the AD and is produced from proteins, urea and nucleic acids degradation. Although TAN is an important nutrient for microbial growth, it is inhibitory at concentrations between 1500 and 3000 mgTAN/L and pH 7.4–7.6, and it is toxic for biomass at concentrations over 3000 mg TAN/L [32]. Free ammonia (NH_3) inhibits the methanogenic process by increasing the maintenance energy requirement, affecting the intracellular pH, depleting the intracellular potassium and inhibiting specific enzyme reactions, principally of archaea populations [33].

Ammonium inhibition at large-scale AD leads to serious economic and operational problems. In fact, many full-scale anaerobic digesters operate in an ammonia-induced "inhibited steady-state," with up to 30% losses of potential methane production yield [1]. Finally, an effluent with a high concentration of ammonia requires treatment before its final disposition, which can be possible with biological treatment or chemical treatment.

Table 1 was divided into two sections: the first one includes raw effluent from industry, which have high content of organic matter and/or nitrogen, whereas the second section groups include anaerobic digestion effluents, which have less organic matter content and more nitrogen content than the respective raw effluent. Optoelectronic industrial wastewater seems to be an ideal substrate for autotrophic processes such as PN-A due to its low COD content; nevertheless, it lacks of essential trace elements [12]. Then, the addition of trace elements is needed to be able to perform a biological treatment. Cokes wastewater has been considered the most toxic one since it contains toxic compounds such as phenols, polynuclear aromatic hydrocarbons, heterocyclic compounds containing nitrogen, oxygen and sulfur, cyanides, thiocyanate and ammonia, nevertheless to employ biological treatment is feasible [18]. Finally, it has been demonstrated that most of effluents presents in **Table 1**, such as poultry manure, slaughterhouse, fish canning, fish industry, cheese whey, etc., can be treated with biological process.

3. Partial nitrification and Anammox process

Autotrophic nitrogen removal technology is used without organic matter, and it is considered the best sustainable treatment for rich nitrogen wastewater. Anaerobic ammonia-oxidizing bacteria (anAOB) are responsible for the anaerobic ammonium oxidation (Anammox) process [34]. Anammox reaction consists of the ammonium oxidation using nitrite as an electron acceptor (Eq. (1)) [35]. Nitrite can be obtained from nitritation (oxidation of ammonium to nitrite) by aerobic ammonia-oxidizing bacteria (aerAOB) (Eq. (2)) [4, 8]. The PN-A process has proved to be an efficient nitrogen removal technology (Eq. (3)) [36].

$$NH_4^+ + 1.146NO_2^- + 0.071HCO_3^- + 0.057H^+ \rightarrow 0.986\,N_2 + 2.02\,H_2O + 0.071CH_{1.74}O_{0.031}N_{0.15} + 0.161NO_3^- \quad (1)$$

$$\mathrm{NH_4^+ + 0.0743\,HCO_3^- + 1.404\,O_2 \rightarrow 0.985\,NO_2^- + 0.0149\,C_5H_7O_2N + 1.911\,H^+ + 1.03\,H_2O} \quad (2)$$

http://dx.doi.org/10.5772/68054

$$NH_4^+ + 0.804\ O_2 + 0.071HCO_3^- \rightarrow 0.436\,N_2 + 0.111N\,O_3^- + 0.009\,C_5H_7O_2N + 0.028CH_2O_{0.5}N_{0.15} + 1.038\,H^+ + 1.46\ H_2O \tag{3}$$

The kinetic parameters of the bacterial groups responsible of the process are present in **Table 2**. Different configurations have been designed to allow a properly balanced process considering the kinetic parameters such as: (1) the low duplication time of aerAOB and anAOB is an advantage because of the low sludge production and in turn is a disadvantage because a high biomass retention is needed [8]. (2) The Anammox activity is temporarily inhibited with dissolved oxygen (DO) at values higher than 0.032 mg/L, but oxygen can be consumed by aerAOB when working with a one-stage system. Then, a correct control of DO is needed. (3) Nitrite oxidizing bacteria (NOB) are an undesirable microorganism, typically present in the process. NOB compete for oxygen with aerAOB and for nitrite

Parameter	Symbol	Value	Unity	Reference
aerAOB				
Maximum growth rate	*Umax aerAOB*	1.36	1/d	[29]
Oxygen saturation coefficient	$k_{O_2}^{aerAOB}$	0.3	g O_2/m^3	[29]
ammonia saturation coefficient	$k_{NH_3}^{aerAOB}$	1.1	g N/m^3	[29]
Decay rate	b_{aerAOB}	0.068	1/d	[38]
NOB				
Maximum growth rate	*Umax NOB*	0.79	1/d	[29]
Oxygen saturation coefficient	$k_{O_2}^{NOB}$	1.1	g O_2/m^3	[29]
ammonia saturation coefficient	$k_{NH_3}^{NOB}$	0.51	g N/m^3	[29]
Decay rate	b_{NOB}	0.04	1/d	[38]
anAOB				
Maximum growth rate	*Umax anNOB*	0.052	1/d	[39]
ammonia saturation coefficient	$k_{NH_3}^{anAOB}$	0.03	g N/m^3	[38]
Nitrite saturation coefficient	$k_{HNO_2}^{anAOB}$	0.005	g N/m^3	[38]
Oxygen inhibition coefficient	$k_{O_2}^{anAOB}$	0.01	g O_2/m^3	[40]
Decay rate	b_{anAOB}	0.0026	1/d	[38]

aerAOB, aerobic ammonia-oxidizing bacteria; NOB, nitrite-oxidizing bacteria; anAOB, anaerobic ammonia-oxidizing bacteria.

Table 2. Kinetic parameters of the partial nitrification and Anammox process.

with anAOB. Thus, the NOB suppression of the system is a priority step to reach high efficiencies in the process. (4) Organic matter can inhibit the entire process because of the fast development of heterotrophic bacteria (HB), which competes for oxygen with aerAOB and for living space with anAOB. (5) Finally, different environmental conditions such as temperature, concentrations of free nitrous acid and free ammonia control the process efficiency [8, 37].

As a conclusion, the success of the process is dominated by two great premises: the type of operation strategies (two or one stage and type of reactor) and the environmental conditions related to the inhibition or process optimization.

3.1. Operation strategies: number of stages and type of reactor

The application of the coupled processes of partial nitrification and Anammox can be performed in two different units or in a single one. The first experience with a full-scale Anammox—two stages process was in the Rotterdam wastewater treatment plant (WWTP) in 2002. The Anammox reactor was coupled to a previous Single reactor system for High Ammonium Removal Over Nitrite (SHARON®) to remove the nitrogen from a side stream [41]. Thus, the first large-scale proposal for the autotrophic removal of nitrogen was composed of two stages: partial nitrification (PN) and anammox (A). SHARON® was designed to produce a partial nitrification by controlling the effluent composition (equal concentration of ammonia and nitrite), the temperature (near to 30°C), the solid retention time (SRT) equal to the HRT, short HRT (1 day) and the pH value through DO concentration. With those strategies, the growth of aerAOB is favored over that of NOB [3, 41]. The NOB suppression has been one of the main challenges of the PN-A systems. Some of strategies are as follows: (1) increasing free ammonia concentration working at high pH values and thus limiting the growth of NOB due to their higher sensitivity to free ammonia than aerAOB [42], (2) decreasing the dissolved oxygen concentration due to the low oxygen affinity of NOB compared to aerAOB [29], (3) operating at temperatures above 25°C since the maximum specific growth rate of aerAOB will be higher than that of NOB at these conditions.

The advantages of a two stages PN-A process are as follows: (1) the organic material can be depleted in the first stage avoiding the anAOB inhibition, (2) all inhibition strategies of NOB can be applied in the first stage, (3) there is no risk of oxygen inhibition of anAOB, and (4) in summary, the aerobic and anaerobic metabolisms can be optimized separately [43].

Operation parameters for two stages processes have been extensively reported. Values for SRT, HRT and mixed liquor suspended solids (MLSSs) ranges in the first partial nitrification unit are 1–13 d, 1–1.25 d and 0.27–20 g MLSS/L, respectively. On the other hand, the second units (anammox reactors) show operation parameter such as HRT and MLSS of 0.5–1.7 d and 0.2–35 g MLSS/L, respectively; SRT is a parameter little measured, because the systems are oriented in retaining the greater quantity of biomass [36, 44–46]. Most nitrogen load rate (NLR) and nitrogen removal efficiency ranges of the combined systems are 0.35–1.2 kg N/ m^3 d and 72–89%, respectively [36, 44–46]. Nevertheless, the highest nitrogen load and removal efficiency have been reported for the Rotterdam anammox reactor with more than 6.5 years of operation period with a high granular biomass concentration of 35 g MLSS/L

[46]. This two stages process has a common NLR and efficiency of 7 kg N/m^3 d and 95%, respectively [46].

Otherwise, the one-stage operation parameters such as SRT, HRT and MLSS are 15–40 d, 0.075–4 d and 2–3.5 g MLSS/L, respectively [12, 36, 47, 48]. In addition, the NLR and nitrogen removal efficiency of this one step process are 0.46–1.4 kg N/m^3 d and 50–89%, respectively [12, 36, 47, 48]. Clearly, greater NLR and efficiencies values are expected in two-stage systems. Despite these advantages of the two-step configuration, 88% of all plants are operated as single-stage systems [46]. The one-stage systems have advantages such as: (1) continuous consumption of nitrite avoiding inhibitions in both aerAOB and anAOB, (2) smaller operational units are needed, (3) simplification of the operation control and (4) lower N_2O emissions compared to two stages systems [49]. In a one-stage reactor, the process has been registered with different names; CANON: Completely Autotrophic Nitrogen removal Over Nitrite process [50]; ELAN: Spanish acronym for ELiminación Autotrófica de Nitrógeno-(autotrophic nitrogen removal) [51]; DEMON: DE-amMONnification [52, 53]; ClearGreen: Cyclic Low Energy Ammonium Removal [46]; NAS: New Activated Sludge [54], OLAND: Oxygen-Limited Autotrophic Nitrification–Denitrification [55]; SNAD: Simultaneous partial Nitrification, Anammox and Denitrification [56].

Some one-stage characteristics are as follows: (1) CANON process is based on the control of parameters such as pH, DO, and redox potential; aeration and shear stress applied to biomass allows the development of a granular biomass aerAOB (external zone) and anAOB (internal zone). (2) ELAN system is operated in cycles of 3 or 6 h where the feeding to the reactor and the aeration s continuous during the most of the time cycle (90–95%). Short periods of settling time are used to allow the washout of NOB flocculent biomass. (3) DEMON is a system with a hydrocyclone that keeps the granular biomass in the reactor and eliminates the small flocculent biomass. (4) ClearGreen is operated with a three-period cycle: At period 1 feeding, mixing, aerobic period and anoxic periods carried out. At period 2 is a settling period and at period 3 withdrawal occurs. Due to the nitrate removal during the anoxic periods nitrogen removal reaches 90%. (5) NAS is in an active sludge with a portion of anAOB. This process shows the combination of batch-fed partial nitritation, anammox, denitrification and nitrification reactors in a four-stage configuration plant with internal recycling lines. (6) OLAND is carried out on biodiscs under microaerobic conditions with coexistence of aerAOB and anAOB.

FISH analyses [55, 57] revealed that anaerobic ammonium oxidation in all aforementioned processes is performed by anAOB. In addition, the coupled reactions of PN-A leave 11% of residual nitrogen in the form of nitrate due to the reaction stoichiometry (see Eq. (4)); thus, in the presence of organic carbon, the remaining nitrate can be used by denitrifying bacteria as an electron acceptor, improving the N removal efficiency. This new process is known as Simultaneous Nitrification, Anammox and Denitrification (SNAD) process [56].

Beyond the regime used (sequencing batch or continuous reactors), all reactor designs for PN-A pursue to retain the biomass in the system due to their long duplication time. Initially, the Anammox process was operated in continuous biofilm reactors [58, 59]. In order to improve the biomass retention and the stability of the process, the sequencing batch reactor

(SBR) has been extensively used [49, 60] where mixing was achieved either by mechanical stirring or by gas flow stirring. More than 50% of all PN-A industrial installations are SBR [46]. In SBR or airlift reactors with suspended biomass, the biomass settling properties determine the retention and are related to the microbial aggregate morphology as floc or granule and size. Granules are defined as compact and dense aggregates with an approximately spherical external appearance that do not coagulate under decreased hydrodynamic shear conditions and settle significantly faster than flocs [61]. In terms of physical properties, large granules are preferable for suspended-growth applications.

Granular biomass allows the development of aerAOB at the external layers of the granule, while anAOB can grow in the anoxic core of granule, but still close to the bulk liquid and to the layer of the aerAOB [51]. In an one-stage PN-A processes, the aggregates sizes not only influence settling properties but also affect the proportion of microbial nitrite production and consumption; low aerAOB activity and high anAOB activity have been observed in large aggregates [62]. Better performances in terms of cost efficiency have been obtained when granular systems were used in a PN-A process [49].

In summary, the most used configuration for PN-A process is a one-stage reactor, mainly because of the lower investment cost compared to a two-stage reactor and for its easy operation. The type of biomass structure depends mainly on the reactor design or regime, where best results have been observed with granular biomass.

3.2. Environmental conditions

The PN-A process is very sensitive to oxygen, temperature, and concentrations of organic matter, free nitrous acid and free ammonia. The anoxic recovery of an autotrophic process is a typical answer to the temporal inhibition of Anammox biomass when DO is near to 0.032 mg O_2/L [40]. Otherwise, when a one-stage PN-A system is operated, the aerobic community such as aerAOB, heterotrophic biomass or even NOB can remove the oxygen before reaching the anAOB cell [63]. As a counterpart, during this symbiosis, anAOB can consume the NO_2, which is toxic for all bacterial populations in the consortium [8]. The PN-A process saves aeration costs because only half of the ammonium needs to be oxidized to nitrite (partial nitritation). Thus, the avoidance of high DO concentrations prevents the growth of NOB and avoids the inhibition of anAOB. NOB has lower affinity for oxygen than aerAOB, and it competes for nitrite with anAOB [8, 64].

Indeed, all these assumptions led to the first start-up strategies of the PN-A process, which were focused on acclimation to low oxygen concentrations. ELAN® and Cleargreen® started their process with DO concentrations below 0.5 and 0.8 mgO_2/L, respectively [46]. In addition, OLAND® process and DEMON® processes started with DO below 0.65 and 0.3 mgO_2/L, respectively [46]. Nevertheless, several authors have proposed to start-up with high oxygen concentrations, such as 1 [62], 4.6 [6] and even 6.6 mgO_2/L [5]. The development of a strong nitrifying layer, to increase the protection of anAOB as well as to increase the granular biomass concentration is the main arguments for a high DO concentration at the start-up [51]. However, a higher DO concentration means a more expensive operation. Also, with more oxygen the granular diameter increases, by one side, this leads to a larger sedimentation capacity, but on the other side, with a diameter above 2.20 mm granules floatation may occur, which hinders the operation [8].

Otherwise, nitrate build-up has been reported in 50% of the large-scale plants, this means that the DO concentration control not always provides a good correlation with the nitrogen removal [46]. Higher oxygen concentration and a small nitrifying layer (due to a low oxygen start-up strategy) lead to an oxygen penetration and NOB activity in the core of the granular biomass. Finally, the type of oxygen strategies used to start up a PN-A process is very important because it affects the granules properties.

Temperature is currently the most investigated parameter. The main aim is to introduce the PN-A process to the mainstream of WWTP, and this innovation will open new possibilities in the design of energy production processes [65]. The Anammox reaction has been assayed at incubation temperature between 6 and 43°C [8]. The slope of activity drops quickly after temperature below 20°C [65]. To understand the influence of temperature on the Anammox activity, it is necessary to understand its influence on the activation energy. The activation energy of anAOB is similar to aerAOB (63–72 kJ/mol) [66]. A correct determination of the effect of the temperature on PN-A process considers different temperature coefficients depending on the experimental range and on the biomass history [65]. Unlike other biological processes, the Arrhenius equation considers different slopes for different temperature ranges. On the other hand, acclimated biomass to lower temperatures presents higher specific rates with a major effect on anAOB biomass compared to the aerAOB biomass [65]. Consequently, the temperature effect increases at lower temperatures, but the importance of this effect is closely related to the biomass specie.

When both PN and Anammox processes are carried out in one stage in the presence of organic matter, the development of heterotrophic bacteria (HB) can destabilize the nitrogen removal process. HB have higher growth rates than autotrophic bacteria and thus, competing for living space and substrates. Moreover, HB outcompete aerAOB and anAOB for oxygen and nitrite, respectively [67, 68]. Nevertheless, if suitable operational conditions and inlet $COD_{biodegradable}/N$ ratios are provided, balanced activities among aerAOB, anAOB and HB can be achieved maintaining a high nitrogen removal efficiency [37]. Stoichiometrically, coupled reactions of partial nitrification and Anammox are capable of removing maximum of 89% of ammonium, leaving the remaining 11% of nitrogen in the form of nitrate. In the presence of organic matter, the remaining nitrate can be used by HB as an electron acceptor for the oxidation of organic carbon approaching the theoretical removal of 100% of nitrogen by the combined action of these three bacterial groups. This trabajofinalizado new process has been called SNAD process. Since its appearance in 2009 [56], the number of published articles of SNAD has grown compared with other N removal processes [67, 68]. The first difficulty of the system is the organic load, since an excess of COD destabilizes the bacterial consortium. Generally, the inlet COD/N ratio reported in the literature takes into account the total COD; however, only the biodegradable fraction of organic matter should be counted because it is the available substrate for heterotrophic growth. Most reported SNAD process working at $COD_{biodegradable}/N$ ratios lower than 0.7 have shown good performances [37].

The inhibition by free nitrous acid and free ammonia concentrations is influenced by pH. Free nitrous acid effect is the most dramatic; indeed, aerAOB catabolic processes present 50% inhibition at 0.40–0.63 mg H-NO_2-N/L under aerobic condition and anabolic process presented complete inhibition at 0.40 mg H-NO_2-N/L under aerobic condition [69]. These concentrations decrease for NOB population. Under aerobic condition, NOB anabolic process has presented completely inhibition at 0.02 mg H-NO_2-N/L and did not present any inhibition for catabolic process up to 0.024 mg H-NO_2-N/L [69]. On other hand, NOB inhibition at free ammonia occurs with concentrations below 1 mg N/L, whereas aerAOB showed inhibition above 16 mg/L [42, 69].

Finally, the operational parameters such as oxygen concentration, temperature, organic matter in the influent, free nitrous acid and free ammonia concentrations are essential for the correct performance of the process.

3.3. Removal of organic matter and nitrogen species (REMON)

The **REM**oval of **O**rganic matter and **N**itrogen species (called REMON) process is based on the sequential and parallel reactions of PN-A (Eq. (3)) and denitrification (Eq. (4)).

$$NO_3^- + 1.08CH_3OH + 0.24H_2CO_3 \rightarrow 0.47N_2 + 0.056C_5H_7O_2N + 1.68H_2O + HCO_3^- \quad (4)$$

The REMON process has been validated in a bubble column reactor (BCR) in a continuous regimen. BCR was selected because the mixing is performed sparging recirculated gas, and this reactor configuration requires less energy than mechanical stirring [64]. Also, the process has been assayed in SBR with mechanical agitation, but the performance is very sensitive to the type of agitation and to oxygen modifications [70]. In the BCR, the upper section had a three-phase separator for granular and the flocculent biomass separation [64].

REMON shows same reaction of SNAD process, but the operational strategies are different. The NOB suppression, granular biomass selection during the continuous process and the good response to high organic matter concentration are main goals of the REMON process.

The NOB inhibition in a PN-A process has been widely studied. Different technologies based on the PN-A process consider strategies such as control of SRT, pH, DO limitation, aeration intensity, redox potential and concentrations of free ammonia [50, 52, 53, 55]. The REMON process has evaluated single-parameter strategies only with oxygen limitation [64]. The best nitrogen removal was 75.36% using a DO of 0.2 mgO_2/L. In addition, at this DO, molecular analyses demonstrated that the NOB group was the most abundant bacteria (**Figure 1**). Understanding the inhibition as the loss of metabolic activity and the suppression as bacterial lysis, the oxygen limitation promoted NOB inhibition without NOB suppression. Thus, the design of a NOB suppression strategy prior to the adaptation of the PN-A biomass to organic matter is crucial [64].

Thus, for NOB suppression, a multiparametric strategy was sized [36]. In order to achieve a robust REMON process capable of tolerate the addition of organic matter, a three stages stabilization strategy was implemented: NOB suppression by free ammonia overload with oxygen limitation, recovery of ammonium oxidizing activity and promotion of aerAOB growth, and

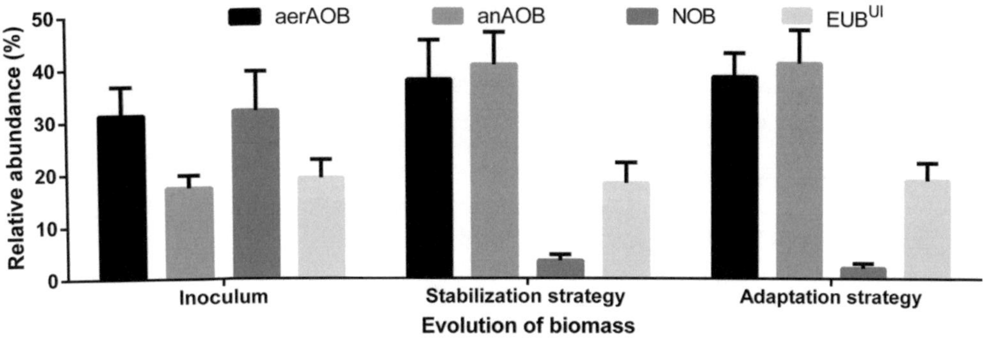

Figure 1. Microbiological characterization of the granular PN-A Inoculum biomass and PN-A biomass during different stages of adaptation. Aerobic ammonia-oxidizing bacteria (aerAOB), anaerobic ammonia-oxidizing bacteria (anAOB), nitrite-oxidizing bacteria (NOB) and unidentified eubacteria (EUB^{UI}) [71].

finally, DO decrease to induce anAOB activity recovery. On the one hand, the FISH analysis confirms a strong decrease of the NOB group in the granular biomass and at the end of the stabilization period, the relative abundance of aerAOB, anAOB, NOB and unidentified eubacteria (EUB^{UI}) was 37.88, 40.67, 3.34 and 18.11%, respectively. On the other hand, the relative abundances of the inoculum were 31.3, 17.4, 32.1 and 19.2%, respectively (**Figure 1**). These results revealed a decrease of nearly 90% of NOB abundance, which support the effectiveness of the start-up implemented strategy. These results agree with the strategies described for the start-up of the PN-A process [50, 52, 53, 55, 64], since so far the literature only reports the inhibition of NOB without considering a bioprospecting of the bacterial consortium in the biomass that will ensure the suppression of undesirables species [71]. Then, an adaptation strategy was performed in four steps, corresponding to different increasing feeding ratios of 25, 50, 75 and 100% (v/v) of anaerobic digester/total substrate ratio (mixed anaerobic digester and synthetic substrate). The aim of a gradual adaptation of the REMON biomass to the organic matter was to avoid an excessive growth of heterotrophic flocculent biomass through the control of the SRT with a slight modification of the separation system. The proposed control was gradual in order to maintain a denitrifying activity on the reactor, prevented also in other systems such as CANON [50]. In addition, the growth of denitrifying bacteria over the granular biomass should be avoided in order to increase the process efficiency. This was successfully accomplished by using a HRT larger than the inverse of the specific growth rate of the heterotrophic bacteria, about 4 h to prevent forming granules and to promote its free floc state [37].

During the experimental work, the biomass concentration in the reactor was 6.5 g VSS/L, and the SRT was 10 d. After a slight modification of the separation system, the biomass concentration on the reactor decreased to 3.5 g VSS/L, and the SRT was 5.4 d. Moreover, the washed-out biomass mainly corresponded to flocs, achieving good granular biomass retention. The control of HRT and SRT allowed the suppression of NOB biomass (**Figure 1**) and the development of denitrifying biomass in the reactor. Summarizing, a greater efficiency was achieved when the adaptation was completed (100%, v/v real substrate), and a maximum of 91.68% total nitrogen removal was reached with a COD/N ratio of 2.63 (organic load of 864 mg COD/L d^{-})

(**Figure 2**). However, for high COD/N ratios, an effective biomass separation system in the SNAD reactor is essential for both the outflow of suspended biomass and the retention of granules. A bad separation system design can lead to a reactor clogging or a fully biomass retention, and the process will collapse [71].

The effect of shear stress on the granular biomass of a REMON continuous reactor system fed with digested poultry manure has been studied [72]. The start-up was carried out in a continuously fed granular BCR. The BCR was stabilized with synthetic substrate and then adapted to digested poultry manure until reaching a NLR of 0.4 g N/L d. After adaptation, the applied power in the BCR was increased from 8.43 to 15.72 W/m^3. The biomass was characterized physicochemical and molecularly. During the increase of the shear stress, nitrogen removal decreased from 63 to 17%. Relative abundance of aerAOB and anAOB did not show significant differences. However, the specific Anammox and nitrification activities fell 88.54 and 53.10%, respectively (see **Table 3**). In summary, there is an upper limit of the applied agitation power on a granular biomass in a REMON reactor. If this limit is exceeded, a negative effect on the activities of the biomass and in the reactor performance is shown [72].

The different operation parameters of the process have shown some limitation such as: 0.25–1 g SST/L$^-$, 0.13–0.5 g SSV/L, a maximum COD/N ratio of 2.63. With an optimum of 0.7 of $COD_{biodegradable}$/N [37], TAN influent concentration of 0.2–0.8 g N/L with HRT of 4–0.4 d, the NLR assay has been 0.05–1 g TAN/L d$^-$. The removal efficiency of the system is 20–50% of COD, with a nitrogen removal of 80–95% [71].

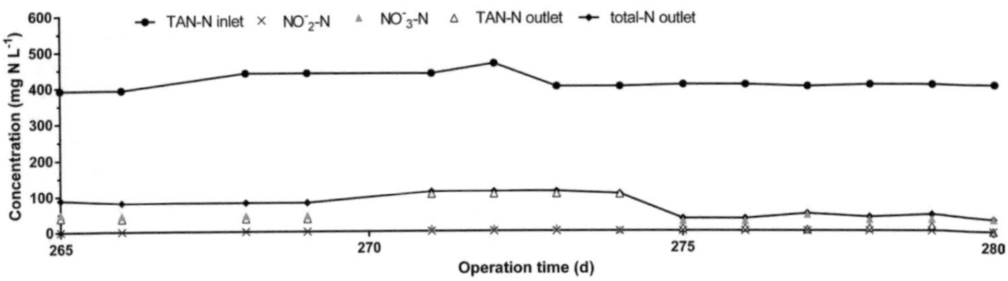

Figure 2. Profile during the adaptation of the PN-A reactor [71].

Shear stress (W/m^3)	EUB [Bacteria/g of biomass]	anAOB [Bacteria/g of biomass]	aerAOB [Bacteria/g of biomass]	NOB [Bacteria/g of biomass]	SNA [g N-NH_4^+/g SSV d]	SAA [g N_2/g SSV d]
8.43	2.86×10^8	1.08×10^8	1.02×10^8	5.18×10^6	0.314	0.113
12.07	2.76×10^8	9.98×10^7	9.13×10^7	7.97×10^6	0.218	0.042
15.72	2.26×10^8	7.02×10^7	8.08×10^7	2.79×10^6	0.036	0.053

EUB, eubacteria; anAOB, anaerobic ammonia-oxidizing bacteria; aerAOB, aerobic ammonia-oxidizing bacteria; NOB, nitrite-oxidizing bacteria; SNA, specific nitrification activity; SAA, specific Anammox activity.

Table 3. Evaluation of the effects of shear stress in REMON system [72].

As a conclusion, REMON is a novel system that optimizes the removal of organic matter and nitrogen species considering strategies that allow NOB suppression and a correct balance between denitrifying bacteria and anAOB.

4. Anaerobic digestion optimization with nitrogen removal: coupled processes

A coupled process prototype at bench scale for the treatment of nitrogen rich wastewaters was developed; the stepwise process has been validated using poultry manure [73]. The first stage comprises the AD of the substrate, where the poultry manure is diluted in order to decrease the ammonium concentration of the substrate to avoid a large inhibition of the methane production. Best results were obtained with three configurations of AD: (1) up flow Anaerobic Sludge Blanket (UASB), (2) thermal pre-treatment with UASB and (3) two stages anaerobic process with a mixed flow reactor (hydrolytic stage) and a UASB (methanogenic stage). In the first step, the diluted manure is anaerobically digested in one of the aforementioned configurations. Most of the organic matter (60–95%) is depleted, and the organic nitrogen of proteins is released in the form of ammonia, reaching high concentrations. Biogas is also generated with a high methane percentage (50–75%). A small fraction of the stabilized solid and an effluent with a remnant organic matter measured as COD is obtained at the outlet stream of the AD. In the second step, the ammonia is removed using a REMON reactor. This reactor generates a warm ammonia free effluent. From the outlet stream, a portion is recirculated to the entrance of the AD, and as a consequence, the slurry inlet stream of the anaerobic digester is diluted (see **Figure 3**).

In the REMON reactor, the denitrifying bacteria uses COD as an electron donor and reduces the residual nitrate to gaseous nitrogen (denitrification process) in presence of organic matter, allowing a complete nitrogen removal and the elimination of the residual biodegradable organic carbon. The integrated process of aerobic nitrification, anaerobic ammonium oxidation and facultative denitrifying bacteria with oxygen limited conditions has the potential of a nearly complete conversion of ammonia and organic carbon to nitrogen gas and carbon dioxide, respectively [71].

The economic and technical feasibility of a coupled process of AD and REMON using water reuse and energy savings applied to a full-scale poultry manure treatment plant was determined to comply with the Chilean environmental law of wastewaters disposal. The new proposed system is more economical than the nitrification-denitrification orthodox processes and offers 15% less sludge generation. The minimum volume of the AD and REMON reactors did not guarantee the minimum annual cost for the plant; on the contrary, a middle case between the minimum and maximum of an objective function of reactors volumes represents the optimal operation condition [74]. But the power consumption is 89.76 and 192.99% lower when burning and using the produced methane, respectively, which means a return of energy. The water recycle results in fresh water savings of 70% compared to the case without recycling. Moreover, the operating costs are reduced by 46%.

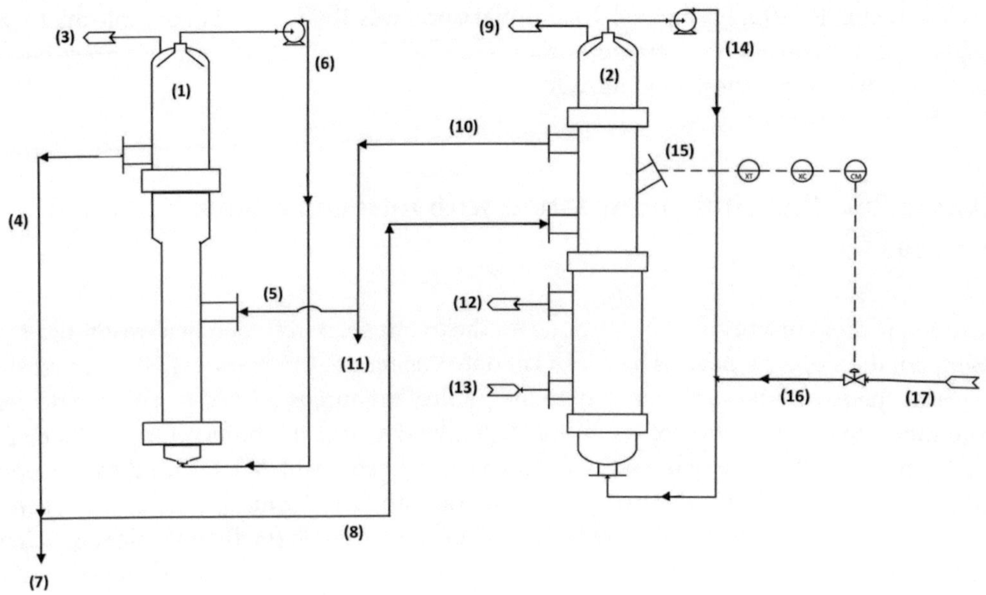

Figure 3. Scheme of coupled processes of anaerobic digester and REMON reactor. (1) anaerobic digester, (2) REMON reactor, (3) overpressure output of anaerobic digester, (4) effluent of anaerobic digester, (5) influent of anaerobic digester, (6) gas recirculation of anaerobic digester, (7) purge of biomass from anaerobic digester, (8) influent of REMON reactor, (9) overpressure output of REMON, (10) effluent of REMON, (11) purge of liquid from REMON, (12) heating water output, (13) heating water input, (14) gas recirculation of REMON, (15) dissolved oxygen [DO] measurement, (16) air make-up, (17) inlet air flow. XT: DO transmitter, XC: DO controller, CM: control module.

5. Conclusions

The correct operation strategies in a biological process (e.g., temperature, nutrient concentrations, bacterial population's interaction and reactor configuration) allow abate high organic carbon and nitrogen concentrations in wastewaters, which poses a problem and changes from a problem to an opportunity. The aforementioned process can be applied to every anaerobic digestion process with inhibitory ammonia concentrations because the need of expensive freshwater can be replaced by recycled treated water with savings of freshwater consumption and operational costs.

Finally, new solutions for ammonia removal using biological treatments reevaluate the technical and economical optimization of the anaerobic digestion projects; the latter were discarded in the past because they showed a negative total annual worth or low biogas potential. Thus, this new process contributes to all different energy matrixes.

Acknowledgements

This chapter was made possible by FONDECYT (Chile) No. 1140491, INNOVA (Chile) No. 12IDL2-13605 and INNOVA (Chile) No. 15VEIID-45613.

Author details

Marlene Roeckel von Bennewitz*, Constanza Beatriz Arriagada Gajardo and Víctor Gabriel Guzmán-Fierro

*Address all correspondence to: mroeckel@udec.cl

Bioengineering Laboratory, Department of Chemical Engineering, Faculty of Engineering, University of Concepción, Concepción, Chile

References

[1] Fotidis IA, Wang H, Fiedel NR, Luo G, Karakashev DB, Angelidaki I. Bioaugmentation as a solution to increase methane production from an ammonia-rich substrate. Environmental Science & Technology. 2014;**48**:7669-7676. DOI:10.1021/es5017075

[2] Batstone DJ, Hülsen T, Mehta CM, Keller J. Platforms for energy and nutrient recovery from domestic wastewater: A review. Chemosphere. 2015;**140**:2-11. DOI:10.1016/j.chemosphere.2014.10.021

[3] van Loosdrecht MCM. Innovative nitrogen removal. In: Biological wastewater treatment-principles, modelling and design. London: IWA; pp. 139-154.

[4] Fux C, Siegrist H. Nitrogen removal from sludge digester liquids by nitrification/denitrification or partial nitritation/anammox: Environmental and economical considerations. Water Science and Technology. 2004;**50**:19-26.

[5] Vázquez-Padín J, Mosquera-Corral A, Campos J, Méndez R, Revsbech N. Microbial community distribution and activity dynamics of granular biomass in a CANON reactor. Water Research. 2010;**44**:4359-4370. DOI:10.1016/j.watres.2010.05.041

[6] Vázquez-Padín JR, Fernández I, Morales N, Campos JL, Mosquera-Corral A, Méndez R. Autotrophic nitrogen removal at low temperature. Water Science and Technology. 2011;**63**:1282-1288. DOI:10.2166/wst.2011.370

[7] Vázquez-Padín JR, Pozo MJ, Jarpa M, Figueroa M, Franco A, Mosquera-Corral A, Campos JL, Méndez R. Treatment of anaerobic sludge digester effluents by the CANON process in an air pulsing SBR.Journal of Hazardous Materials. 2009;**166**:336-341. DOI:http://dx.doi.org/10.1016/j.jhazmat.2008.11.055

[8] Ma B, Wang S, Cao S, Miao Y, Jia F, Du R, Peng Y. Biological nitrogen removal from sewage via anammox: Recent advances. Bioresource Technology. 2016;**200**:981-990. DOI:10.1016/j.biortech.2015.10.074

[9] Aspé E, Marti MC, Roeckel M. Anaerobic treatment of fishery wastewater using a marine sediment inoculum. Water Research. 1997;**31**:2147-2160. DOI:10.1016/S0043-1354(97)00051-1

[10] Carrera J, Vicent T, Lafuente J. Effect of influent COD/N ratio on biological nitrogen removal (BNR) from high-strength ammonium industrial wastewater. Process Biochemistry. 2004;**39**:2035-2041. DOI:10.1016/j.procbio.2003.10.005

[11] Coskun T, Debik E, Demir NM. Treatment of olive mill wastewaters by nanofiltration and reverse osmosis membranes. Desalination. 2010;**259**:65-70. DOI:10.1016/j.desal.2010.04.034

[12] Daverey A, Su S-H, Huang Y-T, Chen S-S, Sung S, Lin J-G. Partial nitrification and anammox process: A method for high strength optoelectronic industrial wastewater treatment. Water Research. 2013;**47**:2929-2937. DOI:http://dx.doi.org/10.1016/j.watres.2013.01.028

[13] Deng LW, Zheng P, Chen ZA. Anaerobic digestion and post-treatment of swine wastewater using IC-SBR process with bypass of raw wastewater. Process Biochemistry. 2006;**41**:965-969. DOI:10.1016/j.procbio.2005.10.022

[14] Farizoglu B, Keskinler B, Yildiz E, Nuhoglu A. Cheese whey treatment performance of an aerobic jet loop membrane bioreactor. Process Biochemistry. 2004;**39**:2283-2291. DOI:10.1016/j.procbio.2003.11.028

[15] Feng JW, Sun YB, Zheng Z, Zhang JB, Li S, Tian YC. Treatment of tannery wastewater by electrocoagulation. Journal of Environmental Sciences. 2007;**19**:1409-1415. DOI:10.1016/S1001-0742(07)60230-7

[16] Gannoun H, Bouallagui H, Okbi A, Sayadi S, Hamdi M. Mesophilic and thermophilic anaerobic digestion of biologically pretreated abattoir wastewaters in an upflow anaerobic filter. Journal of Hazardous Materials. 2009;**170**:263-271. DOI:10.1016/j.jhazmat.2009.04.111

[17] Guo J, Peng Y, Wang S, Zheng Y, Huang H, Wang Z. Long-term effect of dissolved oxygen on partial nitrification performance and microbial community structure. Bioresource Technology. 2009;**100**:2796-2802. DOI:10.1016/j.biortech.2008.12.036

[18] Marañón E, Vázquez I, Rodríguez J, Castrillón L, Fernández Y, López H. Treatment of coke wastewater in a sequential batch reactor (SBR) at pilot plant scale. Bioresource Technology. 2008;**99**:4192-4198. DOI:10.1016/j.biortech.2007.08.081

[19] Sánchez E, Borja R, Weiland P, Travieso L, Martín A. Effect of substrate concentration and temperature on the anaerobic digestion of piggery waste in a tropical climate. Process Biochemistry. 2001;**37**:483-489. DOI:10.1016/S0032-9592(01)00240-0

[20] Tang C-J, Zheng P, Chen T-T, Zhang J-Q, Mahmood Q, Ding S, Chen X-G, Chen J-W, Wu D-T. Enhanced nitrogen removal from pharmaceutical wastewater using SBA-ANAMMOX process. Water Research. 2011;**45**:201-210. DOI:10.1016/j.watres.2010.08.036

[21] Vidal G, Aspé E, Martí MC, Roeckel M. Treatment of recycled wastewaters from fishmeal factory by an anaerobic filter. Biotechnology Letters. 1997;**19**:117-121. DOI:10.1023/A:1018395827115

[22] Wang SG, Liu XW, Gong WX, Gao BY, Zhang DH, Yu HQ. Aerobic granulation with brewery wastewater in a sequencing batch reactor. Bioresource Technology. 2007;**98**:2142-2147. DOI:10.1016/j.biortech.2006.08.018

[23] Wang CC, Lee PH, Kumar M, Huang YT, Sung S, Lin JG. Simultaneous partial nitrification, anaerobic ammonium oxidation and denitrification (SNAD) in a full-scale landfill-leachate treatment plant. Journal of Hazardous Materials. 2010;**175**:622-628. DOI:10.1016/j.jhazmat.2009.10.052

[24] Yang Q, Yang M, Zhang S, Lv W. Treatment of wastewater from a monosodium glutamate manufacturing plant using successive yeast and activated sludge systems. Process Biochemistry. 2005;**40**:2483-2488. DOI:10.1016/j.procbio.2004.09.009

[25] Dosta J, Galí A, Benabdallah El-Hadj T, Macé S, Mata-Álvarez J. Operation and model description of a sequencing batch reactor treating reject water for biological nitrogen removal via nitrite. Bioresource Technology. 2007;**98**:2065-2075. DOI:10.1016/j.biortech.2006.04.033

[26] Carmen F, Anuska MC, Luis CJ, Ramón M. Post-treatment of fish canning effluents by sequential nitrification and autotrophic denitrification processes. Process Biochemistry. 2013;**48**:1368-1374. DOI:10.1016/j.procbio.2013.06.005

[27] Park J, Jin H-F, Lim B-R, Park K-Y, Lee K. Ammonia removal from anaerobic digestion effluent of livestock waste using green alga Scenedesmus sp. Bioresource Technology. 2010;**101**:8649-8657. DOI:10.1016/j.biortech.2010.06.142

[28] Ruiz I, Veiga MC, De Santiago P, Blázquez R. Treatment of slaughterhouse wastewater in a UASB reactor and an anaerobic filter. Bioresource Technology. 1997;**60**:251-258. DOI:10.1016/S0960-8524(97)00020-5

[29] Wiesmann U. Biological nitrogen removal from wastewater. In: Biotechnics/Wastewater. Berlin, Heidelberg: Springer Berlin Heidelberg; pp. 113-154. DOI:10.1007/BFb0008736

[30] Ip YK, Chew SF, Randall DJ. Ammonia toxicity, tolerance, and excretion. Fish Physiology. 2001;**20**:109-148. DOI:10.1016/S1546-5098(01)20005-3

[31] Townsend AR, Howarth RW, Bazzaz FA, Booth MS, Cleveland CC, Collinge SK, Dobson AP, Epstein PR, Holland EA, Keeney DR, Mallin MA, Rogers CA, Wayne P, Wolfe AH. Human health effects of a changing global nitrogen cycle. Frontiers in Ecology and the Environment. 2003;**1**:240-246.

[32] McCarthy PL. Anaerobic waste treatment fundamentals. Part III: Toxic materials and their control. Public Work. 1964;**95**:91-94.

[33] Chen Y, Cheng J, Creamer K. Inhibition of anaerobic digestion process: A review. Bioresource Technology. 2008;**99**:4044-4064. DOI:10.1016/j.biortech.2007.01.057

[34] Mulder A, van de Graaf AA, Robertson LA, Kuenen JG. Anaerobic ammonium oxidation discovered in a denitrifying fluidized bed reactor. FEMS Microbiology Ecology. 1995;**16**:177-184. DOI:10.1111/j.1574-6941.1995.tb00281.x

[35] Lotti T, Kleerebezem R, Lubello C, van Loosdrecht MCM. Physiological and kinetic characterization of a suspended cell anammox culture. Water Research. 2014;**60**:1-14. DOI:http://dx.doi.org/10.1016/j.watres.2014.04.017

[36] del Río ÁV, Fuentes AP, Giustinianovich EA, Gomez JLC, Mosquera-Corral A. Anammox process: Technologies and application to industrial effluents. In: Technologies for the Treatment and Recovery of Nutrients from Industrial Wastewater. IGI Global. pp. 264-289.

[37] Giustinianovich EA, Campos J-L, Roeckel MD. The presence of organic matter during autotrophic nitrogen removal: Problem or opportunity? Sep Purif Technol. 2016;**166**:102-108. DOI:http://dx.doi.org/10.1016/j.seppur.2016.04.012

[38] Volcke EIP, Picioreanu C, De Baets B, van Loosdrecht MCM. Effect of granule size on autotrophic nitrogen removal in a granular sludge reactor. Environmental Technology. 2010;**31**:1271-1280. DOI:10.1080/09593331003702746

[39] Hao X, Heijnen JJ, Van Loosdrecht MCM. Sensitivity analysis of a biofilm model describing a one-stage completely autotrophic nitrogen removal (CANON) process. Biotechnology & Bioengineering. 2001;**77**:266-277. DOI:10.1002/bit.10105

[40] Strous M, Kuenen JG, Jetten MS. Key physiology of anaerobic ammonium oxidation. Applied and Environmental Microbiology. 1999;**65**:3248-3250.

[41] Van Dongen U, Jetten MSM, Van Loosdrecht MCM. The SHARON®-Anammox® process for treatment of ammonium rich wastewater. Water Science and Technology. 2001;**44**:153-160.

[42] Anthonisen AC, Loehr RC, Prakasam TB, Srinath EG. Inhibition of nitrification by ammonia and nitrous acid. Journal of the Water Pollution Control Federation. 1976;**48**:835-852.

[43] Chandra R. Environmental waste management. CRC Press; 2016.

[44] Akgul D, Aktan CK, Yapsakli K, Mertoglu B. Treatment of landfill leachate using UASB-MBR-SHARON-Anammox configuration. Biodegradation. 2013;**24**:399-412. DOI:10.1007/s10532-012-9597-y

[45] Yamamoto T, Takaki K, Koyama T, Furukawa K. Long-term stability of partial nitritation of swine wastewater digester liquor and its subsequent treatment by Anammox. Bioresource Technology. 2008;**99**:6419-6425. DOI:10.1016/j.biortech.2007.11.052

[46] Lackner S, Gilbert EM, Vlaeminck SE, Joss A, Horn H, van Loosdrecht MCM. Full-scale partial nitritation/anammox experiences—An application survey. Water Research. 2014;**55**:292-303. DOI:http://dx.doi.org/10.1016/j.watres.2014.02.032

[47] Wang Y, Wang Y, Wei Y, Chen M. In-situ restoring nitrogen removal for the combined partial nitritation-anammox process deteriorated by nitrate build-up. Biochemical Engineering Journal. 2015;**98**:127-136. DOI:10.1016/j.bej.2015.02.028

[48] De Clippeleir H, Vlaeminck SE, De Wilde F, Daeninck K, Mosquera M, Boeckx P, Verstraete W, Boon N. One-stage partial nitritation/anammox at 15 °C on pretreated sew-

age: Feasibility demonstration at lab-scale. Applied Microbiology and Biotechnology. 2013;**97**:10199-10210. DOI:10.1007/s00253-013-4744-x

[49] Vlaeminck S, De Clippeleir H, Verstraete W. Microbial resource management of one-stage partial nitritation/anammox. Microbial Biotechnology. 2012;**5**:433-448. DOI:10.1111/j.1751-7915.2012.00341.x

[50] Sliekers AO, Derwort N, Gomez JLC, Strous M, Kuenen JG, Jetten MSM. Completely autotrophic nitrogen removal over nitrite in one single reactor. Water Research. 2002;**36**:2475-2482. DOI:http://dx.doi.org/10.1016/S0043-1354(01)00476-6

[51] Morales N, Val del Rio A, Vazquez-Padin JR, Gutierrez R, Fernandez-Gonzalez R, Icaran P, Rogalla F, Campos JL, Mendez R, Mosquera-Corral A. Influence of dissolved oxygen concentration on the start-up of the anammox-based process: ELAN®. Water Science and Technology. 2015;**72**:520-527. DOI:10.2166/wst.2015.233

[52] O'Shaughnessy M, Sizemore J, Musabyimana M, Sanjines P, Murthy S, Wett B, Takács I, Houweling D, Love NG, Pallansch K. Operations and process control of the deammonification (DEMON) process as a sidestream option for nutrient removal. Proceedings of the Water Environment Federation. 2008;**9**:6333-6348. DOI:https://doi.org/10.2175/193864708788809743

[53] Wett B. Development and implementation of a robust deammonification process. Water Science and Technology. 2007;**56**:81-88. DOI:10.2166/wst.2007.611

[54] Desloover J, De Clippeleir H, Boeckx P, Du Laing G, Colsen J, Verstraete W, Vlaeminck SE. Floc-based sequential partial nitritation and anammox at full scale with contrasting N2O emissions. Water Research. 2011;**45**:2811-2821. DOI:http://dx.doi.org/10.1016/j.watres.2011.02.028

[55] Pynaert K, Smets BF, Wyffels S, Beheydt D, Siciliano SD, Verstraete W. Characterization of an autotrophic nitrogen-removing biofilm from a highly loaded lab-scale rotating biological contactor. Applied and Environmental Microbiology. 2003;**69**:3626-3635. DOI:10.1128/AEM.69.6.3626-3635.2003

[56] Chen H, Liu S, Yang F, Xue Y, Wang T. The development of simultaneous partial nitrification, ANAMMOX and denitrification (SNAD) process in a single reactor for nitrogen removal. Bioresource Technology. 2009;**100**:1548-1554. DOI:10.1016/j.biortech.2008.09.003

[57] Helmer C, Tromm C, Hippen A, Rosenwinkel K-H, Seyfried CF, Kunst S. Single stage biological nitrogen removal by nitritation and anaerobic ammonium oxidation in biofilm systems. Water Science and Technology. 2001;**43**:311-320.

[58] Van De Graaf AA, De Bruijn P, Robertson LA, Jetten MSM, Kuenen JG. Autotrophic growth of anaerobic ammonium-oxidizing micro-organisms in a fluidized bed reactor. Microbiology. 1996;**142**:2187-2196. DOI:10.1099/13500872-142-8-2187

[59] Strous M, Van Gerven E, Kuenen J, Jetten M. Effects of aerobic and microaerobic conditions on anaerobic ammonium-oxidizing (anammox) sludge. Applied and Environmental Microbiology. 1997;**63**:2446-2448.

[60] Dapena-Mora A, Campos J, Mosquera-Corral A, Jetten M, Méndez R. Stability of the ANAMMOX process in a gas-lift reactor and a SBR. Journal of Biotechnology. 2004;**110**:159-170. DOI:10.1016/j.jbiotec.2004.02.005

[61] Lemaire R, Webb RI, Yuan Z. Micro-scale observations of the structure of aerobic microbial granules used for the treatment of nutrient-rich industrial wastewater. ISME Journal. 2008;**2**:528-541. DOI:10.1038/ismej.2008.12

[62] Vlaeminck SE, Cloetens LF, Carballa M, Boon N, Verstraete W. Granular biomass capable of partial nitritation and anammox. Water Science and Technology. 2009;**59**:610-617. DOI:10.2166/wst.2009.3Er

[63] Liu S, Yang F, Xue Y, Gong Z, Chen H, Wang T, Su Z. Evaluation of oxygen adaptation and identification of functional bacteria composition for anammox consortium in non-woven biological rotating contactor. Bioresource Technology. 2008;**99**:8273-8279. DOI:http://dx.doi.org/10.1016/j.biortech.2008.03.006

[64] Varas R, Guzmán-Fierro V, Giustinianovich E, Behar J, Fernández K, Roeckel M. Startup and oxygen concentration effects in a continuous granular mixed flow completely autotrophic nitrogen removal over nitrite reactor. Bioresource Technology. 2015;**190**:345-351. DOI:10.1016/j.biortech.2015.04.086

[65] Lotti T, Kleerebezem R, van Loosdrecht MCM. Effect of temperature change on anammox activity. Biotechnology & Bioengineering. 2015;**112**:98-103. DOI:10.1002/bit.25333

[66] Hendrickx TLG, Wang Y, Kampman C, Zeeman G, Temmink H, Buisman CJN. Autotrophic nitrogen removal from low strength waste water at low temperature. Water Research. 2012;**46**:2187-2193. DOI:http://dx.doi.org/10.1016/j.watres.2012.01.037

[67] Zhang Z, Li Y, Chen S, Wang S, Bao X. Simultaneous nitrogen and carbon removal from swine digester liquor by the Canon process and denitrification. Bioresource Technology. 2012;**114**:84-89. DOI:10.1016/j.biortech.2012.03.006

[68] Liang Y, Li D, Zhang X, Zeng H, Yang Z, Zhang J. Microbial characteristics and nitrogen removal of simultaneous partial nitrification, anammox and denitrification (SNAD) process treating low C/N ratio sewage. Bioresource Technology. 2014;**169**:103-109. DOI:10.1016/j.biortech.2014.06.064

[69] Vadivelu VM, Keller J, Yuan Z. Effect of free ammonia on the respiration and growth processes of an enriched Nitrobacter culture. Water Research. 2007;**41**:826-834. DOI:10.1016/j.watres.2006.11.030

[70] Jara P. Comparison of the performance of sequencing batch reactor and a continuous column reactor for the treatment of poultry waste (Comparación del desempeño de un reactor de carga secuencial y un reactor continuo de columna para el tratamiento de residuos avíc). 2016

[71] Arriagada Gajardo C. Adaptation of granular CANON biomass to effluent from an anaerobic digester fed with poultry manure (Adaptación de biomasa granular canon al efluente de un digestor anaerobio alimentado con estiércol avícola). 2015

[72] Almendras Carrillo FA. Evaluation of shear stress on the biomass associated with simultaneous partial nitrification, anammox and denitrification (SNAD) reactor of treatment of digested poultry manure (Evaluación del esfuerzo de corte en biomasa asociada a reactor de nitrificaci. 2016.

[73] Bustos C. Operational validation of the process of anaerobic digestion and simultaneous nitrification, anammox and denitrification (DA-SNAD) with recycle (Validación operacional de un sistema integrado de digestión anaeróbica y proceso simultáneo de nitrificación a). 2016

[74] Alejo-Alvarez L, Guzmán-Fierro V, Fernández K, Roeckel M. Technical and economical optimization of a full-scale poultry manure treatment process: Total ammonia nitrogen balance. Environmental Technology. 2016;**37**:2865-2878. DOI:10.1080/09593330.2016.1167963

Chapter 3

Heterotrophic Nitrification and Aerobic Denitrification by *Alcaligenes faecalis* No. 4

Makoto Shoda

Additional information is available at the end of the chapter

http://dx.doi.org/10.5772/68052

Abstract

Alcaligenes faecalis No. 4 (No. 4) has the ability to carry out the following heterotrophic nitrification and aerobic denitrification, $NH_4^+ \rightarrow NH_2OH \rightarrow N_2O \rightarrow N_2$. Approximately, 40 and 60% of ammonium were converted to N_2 gas and cell mass, respectively. Only a few percent of NO_2^- and NO_3^- were produced from ammonium. After brief explanation of significant properties of No. 4, several examples of application of No. 4 to treat ammonium, especially high-strength ammonium in several wastewaters were presented. The ammonium removal rates in these examples showed several hundredfold higher than those in conventional ammonium treatment method. In wastewater treatment plants, the selection of handling of excess sludge after treatment is a problem to be solved. Some possibilities of utilization of the excess cells of No. 4 in agriculture or in cattle farming were also presented.

Keywords: heterotrophic nitrification, aerobic denitrification, high-strength ammonium, ammonium removal rates, utilization of organic acids, *Alcaligenes faecalis* No. 4

1. Introduction

1.1. Brief review of conventional ammonium removal by autotrophic nitrification and anaerobic denitrification

The oxidation of the ammonium to nitrogen gas is achieved with two step reactions, namely aerobic nitrification and anaerobic denitrification. The most common bacteria responsible for the aerobic nitrification are the autotrophic organisms, such as *Nitrosomonas* and *Nitrobacter*. They obtain energy from the oxidation of ammonia, obtain carbon from CO_2 and use oxygen as the electron acceptor. Many different heterotrophs are responsible for anaerobic denitrification. They use carbon from complex organic compounds, prefer low to zero dissolved oxygen, and use nitrate as the electron acceptor.

Biological removal of ammonium in a conventional treatment system has been conducted using the two reactions. However, this system faces several problems including: (1) an extremely slow nitrification reaction, (2) deterioration of activity against overloading of ammonium and organic matter, (3) strong sensitivity to oxygen limitation, and (4) requirement of two separate reactors for an aerobic process in nitrification and an anaerobic process in denitrification. The low nitrification rates in this process result in the need for long hydraulic retention times or large reactor volumes to accomplish complete NH_4^+removal. Consequently, conventional treatment demands multiple and larger reactors and high capital and operation costs.

Over the past two decades, several new bioprocesses for ammonium removal from municipal and domestic wastewaters have been developed, including: simultaneous nitrification and denitrification, shortcut nitrification and denitrification, aerobic deammonification, complete autotrophic nitrogen removal over nitrite (CANON), oxygen-limited nitrification and denitrification (OLAND), advanced treatments using combination of these process including membrane bioreactors, and cell-immobilization systems. However, these processes also have some potential problems or limitations especially for high-strength ammonium treatment [1].

1.2. Brief review of anammox method

Anammox (anaerobic ammonium oxidation) is a recent understanding on the nitrogen cycle. *Candidatus* "Brocadia anammoxidans" and *Candidatus* "Kuenenia stuttgartiensis" are representative anammox bacteria.

Anammox method consists of partial aerobic nitrification and anaerobic denitrification.

$$NH_4^+ \rightarrow O_2 \rightarrow NO_2^- \text{ and } NH_4^+ \rightarrow NO_2^- \rightarrow NO_3^- + N_2$$

The advantages of this method are: (1) very little sludge production, (2) reduced oxygen supply, and (3) no need to supplement organic carbons, which are related with operating cost problems in conventional ammonium treatment.

Concerning the ammonium removal rates in anammox method, the relatively higher removal rates, 0.96 kg-N/m^3/day in SHARON-anammox process [2], 2.3 kg-N/m^3/day in fluidized bed using synthetic medium [3], and more than 4 kg-N/m^3/day of gel pellets of anammox biomass [4], were reported. The problems of the method are that: (1) sufficient amount of biomass production is time-consuming, (2) the long time for stabilization of the system, (3) difficult quick recovery of the system when the inefficient removal occurred, (4) NO_3^- accumulation, and (5) slow phosphate removal rate.

1.3. Heterotrophic nitrification and aerobic denitrification

Recently, many microorganisms have been found to conduct heterotrophic nitrification and aerobic denitrification. **Table 1** shows representative microorganisms published previously and their removal abilities. These microorganisms have advantages such as (1) procedural simplicity, where nitrification and denitrification can take place simultaneously, (2) less acclimation problems, (3) lesser buffer quantity needed because alkalinity generated during denitrification can partly compensate for the alkalinity consumption in nitrification.

Strains	Initial NH_4^+-N concentrations (mg/l)	NH_4^+-N removal (%)	Temperature (°C)	N_2 production (%)	Carbon sources	Removal rate (NH_4^+-N kg/m³/day)	Ref. no.
Pseudomonas alcaligenes AS-1	35	100	30	60	Acetate	0.015	[14]
Pseudomonas sp. 3–7	75	86	30	n. r.	Succinate	0.75	[13]
Pseudomonas sp. *Rhodoferax ferrireducens*	80	38	10	n. r.	Acetate	0.06	[9]
Agrobacterium sp.LAD9	97	90	30	50	Succinate	0.088	[6]
Rhodococcus sp. CPZ 24	100	85	30	48	Succinate	0.08	[5]
Bacillus subtilis A1	104	65	28	20	Acetate	0.08	[8]
Pseudomonas stutzeri YZN-001	110	95	30	39	Succinate	0.13	[10]
Acinetobacter calcoaceticus HNR	120	96	30	40.2	Glucose	0.46	[12]
Bacillus methylotrophicus L7	140	70	37	n. r.	Succinate	0.05	[11]
Diaphorobacter sp.	212	100	30	n. r.	Citrate	0.05	[7]
Acinetobacter sp.Y1	110	99	30	54	Citrate	0.25	[15]
Acinetobacter junii YB	100	100	37	51	Succinate	0.59	[16]
Marinobacter sp.	242	48	30	n. r.	Succinate	0.24	[17]

n.r.: not reported.

Table 1. Reported strains which have ability of heterotrophic nitrification and aerobic denitrification.

The two mechanisms for heterotrophic nitrification and aerobic denitrification are reported.

(1) $NH_4^+ \rightarrow NH_2OH \rightarrow NO_2^- \rightarrow NO_3^-$ and $NO_3^- \rightarrow NO_2^- \rightarrow N_2O \rightarrow N_2$

Both reactions occur simultaneously [6, 16].

(2) $NH_4^+ \rightarrow NH_2OH \rightarrow N_2O \rightarrow N_2$

Almost no nitrite or nitrate was produced and neither nitrite nor nitrate was utilized as electron accepters [12, 18].

This kind of bacteria may have the potential to overcome the problems inherent in the conventional nitrogen removal process and to realize one-stage nitrogen removal under aerobic conditions.

In **Table 1**, low-strength ammonium in synthetic medium was used and main carbon sources are organic acids. The use of practical wastewater is scarcely reported. *Alcaligenes faecalis* No. 4 (No. 4) we isolated is one of these microorganisms, and No. 4 showed efficient removal ability for high-strength ammonium and significantly higher removal rate. The following sections present the results when No. 4 was applied to practical wastewaters.

2. Characteristics of *Alcaligenes faecalis* No. 4 (No. 4)

2.1. Basic features of No. 4 [18]

2.1.1. Materials and methods

Strain used: *A. faecalis* No. 4 (No. 4) was isolated from sewage sludge as an antagonistic microorganism to plant pathogens [19].

Synthetic medium used: A synthetic medium containing (in units of g/L) 14 K_2HPO_4, 6 KH_2PO_4, 17 trisodium citrate dihydrate, 2 $(NH_4)_2SO_4$, 0.2 $MgSO_4{\cdot}7H_2O$, and 2 mL of trace mineral solution was used for the preculture of No. 4. The trace mineral solution contained the following components (in g/L): 57.1 EDTA (2,2′,2″,2‴-(ethane-1,2-diyldinitrilo)tetra acetic acid)·2Na, 3.9 $ZnSO_4{\cdot}7H_2O$, 7$CaCl_2{\cdot}2H_2O$, 5.1 $MnCl_2{\cdot}4H_2O$, 5.0 $FeSO_4{\cdot}7H_2O$, 1.1 $(NH_4)_6Mo_7O_{24}{\cdot}4H_2O$, 1.6 $CuSO_4{\cdot}5H_2O$, and 1.6 $CoCl_2{\cdot}6H_2O$.

Method: Available carbon sources and available nitrogen sources were surveyed using various carbon and nitrogen materials. Then, the initial ammonium concentration of $(NH_4)_2SO_4$ was fixed and carbon content of citrate was change from C/N ratio 5–20 and optimal C/N ratio was determined. Optimal temperature and pH were determined using synthetic medium. Nitrogen balance was obtained using NO_x analyzer to detect NO and NO_2 in exhaust gas. All experiments were conducted using shaking flasks (100 ml working volume in 500 ml nominal volume of flask).

2.1.2. Results

The following results were obtained.

Available carbon sources: Organic acids (oxalate, citrate, lactate, formate, acetate, propionate, *iso*-butyrate, *n*-butyrate), amino acids, and phenol. No sugars were available.

Available nitrogen sources: Inorganic ammonium salts, peptone, yeast extract, and hydroxylamine. Neither nitrate nor nitrite was utilized.

Optimal C/N ratio: Optimal C/N ratio was 10 when the NH_4^+-N removal rate was the highest and citrate and ammonium were exhausted simultaneously.

Temperature range: 15–37°C. Optimal temperature was 30°C

Initial pH: In the range of 6–8, ammonium removal rate was almost the same.

Nitrogen balance: Nitrogen balance at the initial 1122 mg-N/l is shown in **Table 2**.

The emitted NO was less than 3% of removed NH_4^+-N.

2.2. Verification of N_2 production directly from ammonium by No. 4 [18]

A 15N tracer experiment using $(15NH_4)_2SO_4$ (50% by atomic fraction, Nippon Sanso Co., Ltd.) was carried out to confirm the production of N_2 by No. 4 in an aerated batch culture in the basic medium under C/N = 10 at 30°C. The exhaust gas was directly introduced into the GC/MS (GC 6850, Agilent Technologies, Japan, Ltd.). The change in nitrogen isotope ratio was measured and N_2 production by No. 4 was calculated from the difference between output $^{29}N_2$ and input $^{29}N_2$.

Figure 1 shows temporal changes in N_2 and N_2O concentrations. It was confirmed that No. 4 can convert NH_4+-N to N_2 gas and that N_2 production ratio among denitrified products was about 90%. In conventional denitrification, 20–30% of influent nitrogen was estimated to be emitted as N_2O under high-strength ammonium conditions. In this system, N_2O production was less than 10% of removed ammonium.

2.3. Ammonium removal under high salt condition by No. 4 [20]

No. 4 exhibited the unique feature of removing ammonium under high salt conditions. **Figure 2** shows change in the ammonium concentration in the cultivation of No. 4 in synthetic medium containing 0, 3, and 6% NaCl in shaking flasks. Ammonium removal began after induction periods of 1 day at 3% NaCl and 5 days at 6% NaCl and the ammonium removal rates were similar to those found in the presence of 0% NaCl. Although No. 4 is not osmophilic, the cells

NH_4^+-N concentration		Nitrification products			Intracellular N (ratio)	Stripped NH_3-N (ratio)	NO-N	Denitrified products (ratio)
Initial	**Final**	**NH_2OH-N**	**NO_2^--N**	**NO_3^--N**				
1122	0	0.03	16	13	510 (46%)	41 (4%)	36	584 (52%)

Table 2. Nitrogen balance (units: mg/l) of ammonium removal in shaken flask experiment by *A. faecalis* No. 4 after 93h cultivation [18].

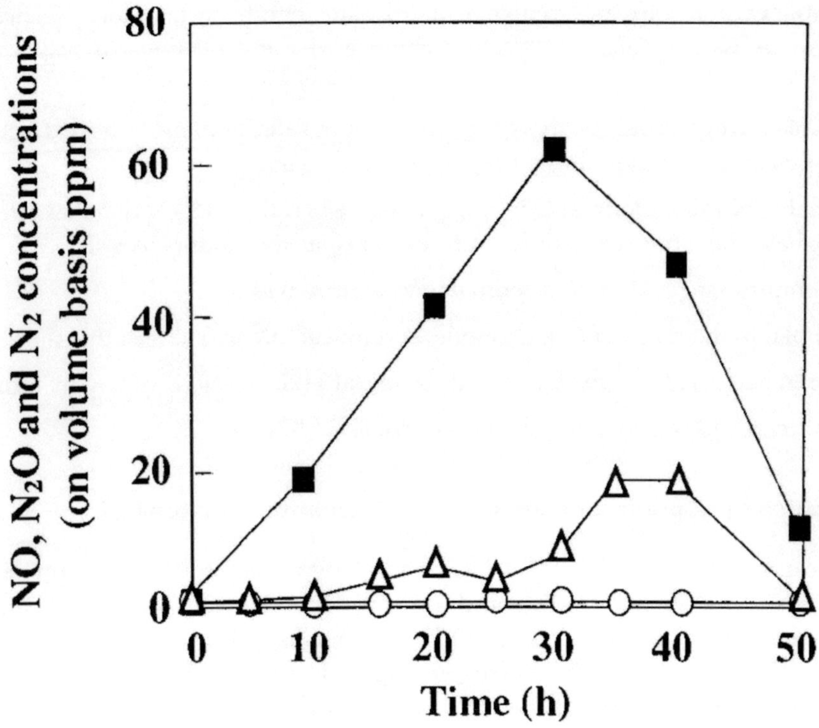

Figure 1. Denitrification characteristics of No. 4 detected by using $(^{15}NH_4)_2SO_4$. Symbols: NO (○), N_2O (△) and N_2 (■) [18].

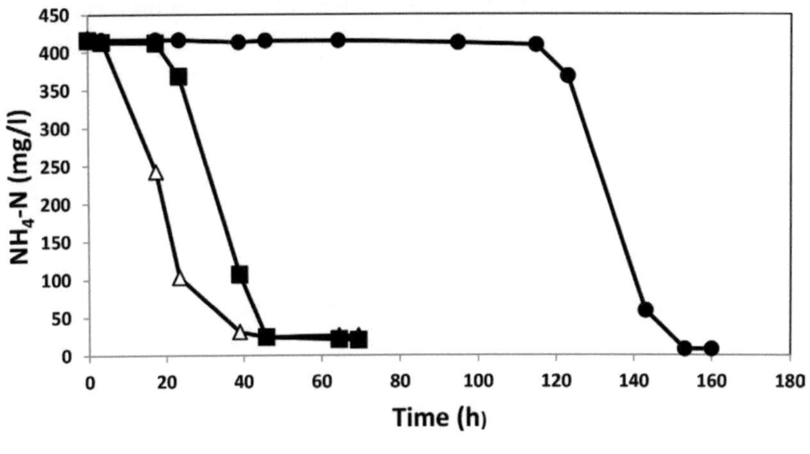

Figure 2. Ammonium removal by No. 4 under 0% NaCl (△), 3% NaCl (■) and 6% NaCl (•) conditions in shaking flasks containing 100 ml of synthetic medium at 30°C [20].

were able to achieve ammonium removal under high saline conditions. In our basic experiment, No. 4 was found to synthesize the osmoprotectant, hydroxyectoine during the lag time when the cells were exposed to high salt concentrations. Because most microorganisms are vulnerable to wastewater with high saline concentrations or high-strength solvents due to the resulting high osmotic pressure, No. 4 is able to effectively remove ammonium under such conditions after a certain acclimation period. Thus, the No. 4 system can remove high-strength ammonium from marine aqua-culture wastewater or fishery processing wastewater.

In the following sections, examples of removal of high-strength ammonium from practical wastewaters are presented.

3. Application of No. 4 to removal of high strength ammonium in various wastewaters

3.1. Crude piggery wastewater [21]

3.1.1. Introduction

Piggery wastewater contains not only high concentrations of nitrogen compounds but also high concentrations of carbon materials. The ammonium concentration reaches up to 1000–3000 mg/l, which is 50–100 times higher than in municipal wastewater. The C/N ratio of the mixture of urine and feces in piggery wastewater is usually in the range of 5–20. Therefore, conventional nitrification using autotrophic bacteria is difficult to apply to such wastewater because nitrification by autotrophic bacteria requires a long retention time of flowing wastewater in the reactor due to the slow growth rates of these autotrophic bacteria. Thus, No. 4 was applied to batch and continuous cultures using solids-free wastewater (referred to as SFW) either alone or supplemented with additional carbon sources such as citrate or feces.

3.1.2. Materials and methods

Microorganisms: The cells of No. 4 were stored in a 25% glycerol solution in vials at −80°C and each vial was used for preculture.

Medium: The synthetic medium described above was used as a preculture. In continuous culture, 500 ml of the preculture was prepared and put into the reactor.

Piggery wastewater: Piggery wastewater was provided by the Kanagawa Prefectural Livestock Industry Research, Kanagawa, Japan. Solids-free wastewater (SFW) was obtained by separating the solids from the raw wastewater containing urine and washing water and feces by centrifugation at 1000 rpm. **Table 3** shows the characteristics of the SFW and mixed wastewater comprised of SFW supplemented with feces (3:1 on a weight basis) (referred to as MW).

Continuous experiments: Continuous treatment of SFW and MW was conducted in a 2.3 l aeration tank at room temperature at the airflow rate of 2.5 l/min. A total of 500 ml of No. 4 culture was mixed with wastewater and open continuous experiments were started by supplying SFW.

	NH_4^+-N	NO_2^--N, NO_3^--N	COD_{Cr}	C/N ratio
*SFW	830–1250	Less than 1	4150–5300	4–5
**MW	1850–1960	Less than 1	13800–14650	7–9

*SFW: Solid-free wastewater (units, mg/l).
**MW: Mixed wastewater (units, mg/l).

Table 3. Characteristics of piggery wastewaters used [21].

3.1.3. Results

3.1.3.1. Batch experiments

SFW or MW was treated with No. 4 in shaking flasks and the removal of NH_4^+-N was measured. In SFW, the addition of citrate was needed mainly due to small amount of carbon in SFW. In MW without addition of carbon, the ammonium removal proceeded smoothly and the maximum ammonium removal rates in SFW with supply of citrate and in MW were 0.7 and 0.66 kg-N/m^3/day, respectively.

3.1.3.2. Continuous experiments

Figure 3 shows the results of SFW and MW treatment in continuous treatment in for 80 days. Initial 10 days, only SFW was supplied and the ammonium removal ratio declined mainly because of lacking of carbon source. Then, citric acid was added and the hydraulic retention time (HRT) was reduced. Consequently, ammonium removal was stabilized to about 80%. Then, the inlet ammonium concentration and citric acid increased gradually and the removal ratio reached almost 100%. From 52 days, instead of supplying citrate, influent NH_4^+-N concentration was increased to 2000 mg/l with addition of feces and HRT was set at 60 h. After 52 days, ammonium removal was high at 100% and outlet of ammonium was less than 20 mg/l. The pH was maintained at 7.4.8 after supply of MW and stripped ammonia from reactor was 2–5% of inlet ammonium concentration. The system was in a steady state. The cells number of No. 4 was measured with L agar plates. The data are summarized in **Table 4**. The nitrogen and carbon balances in the experimental periods are also shown in **Table 4**. After 52 days, feces containing MW were added directly. The denitrified N calculated from inlet ammonium minus the nitrogenous items was 73% and striking result was the high removal of COD. The estimated cell number of No. 4 reached up to 97% of total cells in the samples. The ammonium removal rate, 33 mg-N/l/h corresponded to 0.79 kg-N/m^3/day, which was a few hundred times higher than conventional treatment methods. In diluted and digested piggery wastewater at C/N = 1, 64 mg/l/h removal rate was reported [22]. However, in the present study, No. 4 provided suitable to treat undiluted piggery wastewater with C/N ratio of 10, yielding removal rate of 33 mg-N/l/h (0.79 kg-N/m^3/day).

3.2. Anaerobically digested sludge [20]

3.2.1. Introduction

Due to recent trends of limiting fossil energy consumption, sustainable methods of energy production including methane production in anaerobic digestion or bioethanol production have

http://dx.doi.org/10.5772/68052

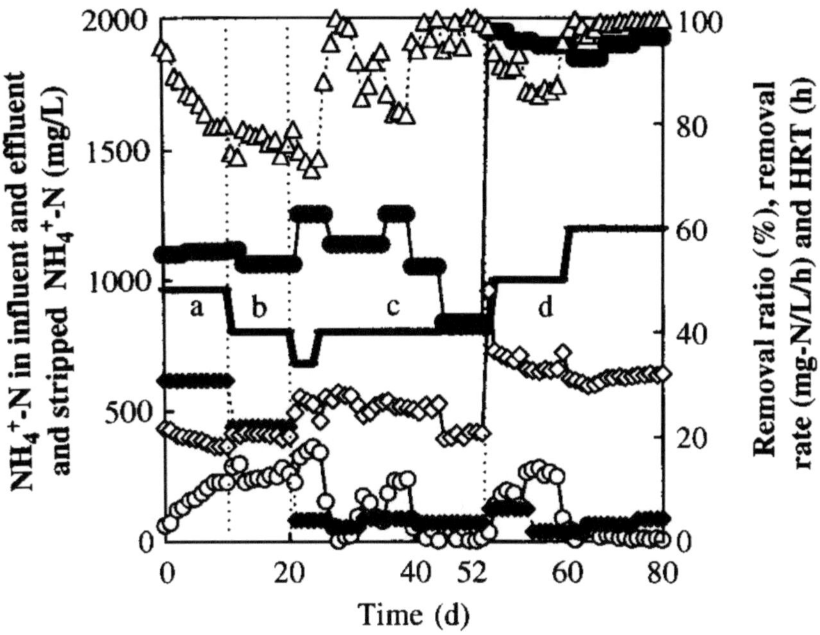

Figure 3. Ammonium concentration in influent (●) and effluent (○), stripped ammonia (◆), removal rate of ammonium (◇), removal ratio of ammonium (△) and hydraulic retention time (HRT) (solid line) in the continuous treatment of solid-free piggery wastewater (before 52 days) and mixed wastewater (after 52 days) by No. 4 [21].

Items	Operation periods	
	21–51 days	**52–80days**
Load of NH_4^+-N (mg/l/day)	670	837
Influent NH_4^+-N (mg/l)	1084	1901
Effluent NH_4^+-N (mg/l)	116	97
Removed NH_4^+-N (mg/l)	968	1804
Intracellular N[a] (mg/l)(%)[b]	232(24)	419(23)
Stripped NH_3-N (mg/l)	74	73
Denitrified N (mg/l)(%)[b]	662(68)	1312(73)
Ammonium removal rate (kg-N/m^3/day)	0.6	0.79
Influent CODcr-C (mg/l)	13,491	12,762
Effluent CODcr-C (mg/l)	1679	342
COD removal ratio (%)	87	97
Cell number of No. 4 (cfu/ml)	1.7 × 10^9	8.8 × 10^9

[a]Calculated by elementary analysis of the dry sludge.
[b]Percentage against removed ammonium nitrogen.

Table 4. The nitrogen balance and carbon change in continuous experiment using No. 4 for treatment of solid free piggery wastewater and mixed wastewater All data were average values in the operation periods [21].

been attracting increasing attention. In anaerobic digestion, livestock waste, municipal garbage, and waste from the food industry are used for the digestion, leading to the production of wastewater containing a high concentration of ammonium. Therefore, the development of an effective method of the wastewater treatment is a crucial factor enabling the production of methane.

In this section, No. 4 was applied to remove high-strength ammonium from digested sludge generated in a municipal anaerobic digestion plant to assess the possibility of efficient biological treatment of the wastewater.

3.2.2. Materials and methods

The reactor used the reactions that were carried out in a small-scale jar fermenter (total volume 1 l, working volume 300 ml). Dissolved oxygen (DO) concentrations and pH values were monitored with a DO sensor and pH sensor inserted into the fermenter. The temperature was controlled at 20 or 30°C. The oxygen transfer coefficients, k_La, were varied by changing the agitation speed from 300 to 700 rpm at a constant air supply rate of 300 ml/min.

Experimental material: The digested sludge was supplied by Yokohama Municipal Sewage Center, Yokohama, Japan, where the excess municipal dehydrated activated sludge was digested at 37°C in a 6000 ton-scale anaerobic digester. The main characteristics of the digested sludge are as follows: pH 7.3, 24 mg/l volatile fatty acids, 2700 mg/l total nitrogen, 1200 mg/l ammonium-nitrogen, 150 mg/l soluble BOD, 1000 mg/l total BOD, 900 mg/l soluble COD, and 20,000 mg/l total COD.

Experimental procedure: The strain No. 4 cells were precultivated in 100 ml synthetic medium in a 500 ml shaking flask at 30°C. The ammonium removal was confirmed in the mixture of 250 ml of digested sludge, 50 ml of strain No. 4 preculture, and 20 g of trisodium citrate dehydrate in the jar fermenter operated at 30°C at an airflow rate of 300 ml/min and at an agitation speed of 700 rpm.

In repeated batch experiments, 50 ml of the preculture of No. 4, 250 ml of the digested sludge, and 20 g of trisodium citrate dihydrate were mixed in the fermenter, and the treatment of the ammonium was conducted. After the ammonium concentration was confirmed to be reduced by more than 90% of the initial concentration, 50 ml of the culture was used for the subsequence treatment by adding a fresh 250 ml of digested sludge and 20 g of trisodium citrate dihydrate.

The optimal C/N ratio for No. 4 was 10, indicating that at this ratio, nitrogen and carbon sources were simultaneously consumed. Based on the ratio, 1 g-N and 10 g-C was balanced and thus 10 g-C corresponded to 38 g of trisodium citrate dihydrate. If no other carbon source existed in the sludge, 38 g of trisodium citrate should be added. In this experiment, 20 g of trisodium citrate dihydrate was arbitrarily chosen by expecting existence of some carbon sources in the sludge.

Analytical method: The ammonium concentration was determined using an ammonium sensor. To determine the number of cells No. 4, the sampled culture was diluted and plated on synthetic agar plates containing the synthetic medium and 1.5% agar, and then the plates were incubated at 30°C for 2 days. As No. 4 grew on the plates significantly faster than other cells indigenous to the digested sludge and that No. 4 exhibited characteristic morphological features, the colonies that appeared on the plates after 2 days were counted as No. 4 cells and the cell concentration was expressed as cells/ml.

3.2.3. Results

3.2.3.1. Ammonium removal in the repeated batch experiment

Figure 4 shows the change in ammonium concentration over times in a repeated batch experiment at 30°C, and **Figure 5** shows the change in the number of No. 4 cells during the same experiment. More than 90% of ammonium was removed within 10–20 h, and the number of No. 4 cells varied between 10^8 and 10^9 cells/ml. The average ammonium removal rate during the experimental period was 2.9 kg-N/m³/day. This value is significantly higher than that in conventional nitrification-denitrification processes and similar to that in an efficient anammox process [3, 4]. Between 169 and 221 h, the operation was stopped and the jar fermenter was left statically at room temperature. When the operation resumed, ammonium removal was observed without any delay, indicating that interrupted operation causes no adverse effect on the activity of No. 4. At 20°C, the average ammonium removal rate decreased to 1.5 kg-N/m³/day.

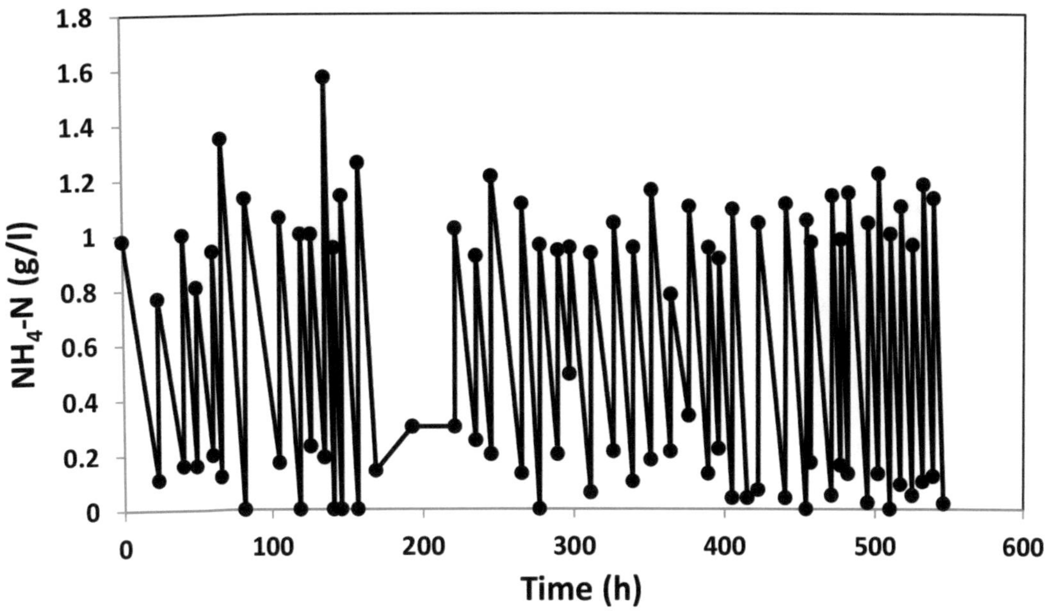

Figure 4. Change in ammonium concentrations of the digested sludge in repeated batch treatment by No. 4. Operation was conducted at 30°C at agitation speed of 700 rpm in a jar fermenter [20].

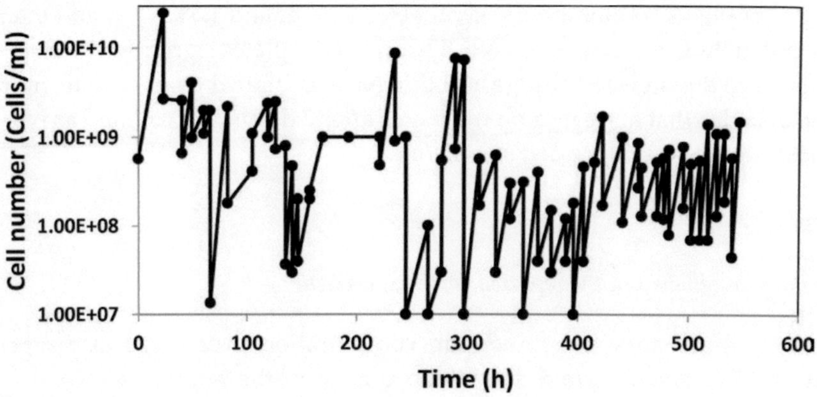

Figure 5. Change in the cell number of No. 4 in the same experiment as shown in **Figure 4** [20].

3.2.3.2. Effect of DO concentration on ammonium removal

In practical operation, DO concentration is related to energy consumption, agitation, air supply, and the activity of No. 4. The effects of changes in the oxygen supply rate on ammonium removal were studied by changing the agitation speed from 700 to 300 rpm. At 700 rpm, the DO concentration was maintained at more than 2 mg/l during the operation, and the ammonium was completely removed within 10 h. However, when the agitation speed was decreased to 400 or 300 rpm, the DO concentration decreased below 1 mg/l, reducing the ammonium removal rate, indicating that the oxygen supply is an important factor for efficient ammonium removal.

3.2.3.3. Ammonium removal under high salt conditions

Strain No. 4 exhibited the unique feature of removal ammonium under high salt conditions as shown in Sections 2 and 3. NaCl was added to the digested sludge to 3%, and repeated batch treatment was conducted at 30°C in a jar fermenter. The results of this experiment are shown in **Figure 6**. The ammonium removal rate reached 3 kg-N/m^3/day at the sixth repeated batch operation after the gradual acclimation of No. 4 to the saline medium.

3.2.3.4. Carbon requirement by No. 4

The experiment performed here included 20 g of trisodium citrate dihydrate. Generally, the C/N ratio of the intracellular components in microorganisms is 10, indicating that 10 units of carbon are used when 1 unit of N is consumed with the C mainly to synthesize cellular materials. In previous experiments, 30–40% of ammonium was reduced to nitrogen gas by No. 4. Assuming a similar level of denitrification in these experiments, 0.6–0.7 g-N/l was used for cell synthesis, indicating that 6–7 g-C/l was required. When 20–30% of carbon is available from the digested sludge, approximately 5 g-C/l should be supplied from outside. As 20 g of trisodium citrate dihydrate contained 16 g-C/l, 6 g of trisodium citrate dihydrate is sufficient to enable the complete removal of 1 g-N/l.

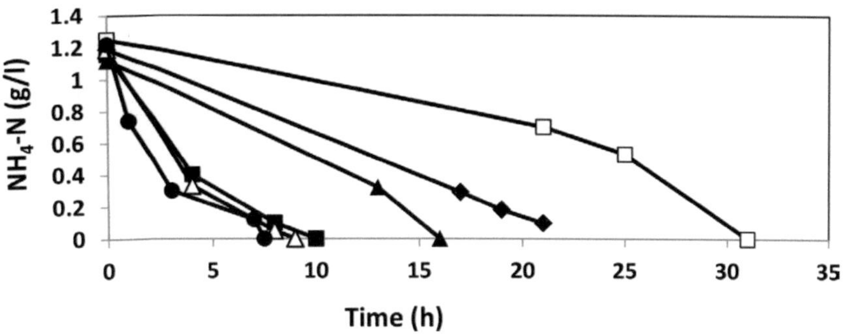

Figure 6. Ammonium removal by No. 4 in the digested sludge containing 3% NaCl by repeated batch experiment. Symbols: 1st (□), 2nd (◆), 3rd (▲), 4th (■), 5th(△) and 6th (•) [20].

Concerning carbon requirement in strain No. 4, the conventional denitrification process using methanol is compared by using the following reaction.

$$NO_3^- + 1.08\,CH_3OH + H^+ \rightarrow 0.065\,C_5H_7NO_2 + 0.47\,N_2 + 0.76\,CO_2 + 2.44\,H_2O \quad (1)$$

The C/N ratio in this reaction is 2. No. 4 process demanded C/N ratio 10. In this point, strain No. 4 process is disadvantageous. However, as a total system, No. 4 process will be advantageous over the conventional process in that no dilution of high strength of wastewater is required, only single reactor with compact size is needed and significantly high removal rate is possible when less expensive carbon sources from waste or unused resources are available. Under these conditions, this system can achieve efficient ammonium removal.

Higher ammonium removal rates have been reported using the anammox method as described in Sections 1 and 2. On the other hand, it is easy to cultivate strain No. 4 in a synthetic medium with a doubling time of 2–3 h. When cultured strain No. 4 cells were stored at 4°C, high activity was maintained for several months, and the cells remained tolerant to high osmotic pressure. The comparison of three methods of ammonium treatment is shown in **Table 8** of Section 5.

In relatively small-scale reactors like this jar fermenter, oxygen supply capacity is lower than those large-scale reactors. The power requirement in wastewater treatment is one of the important factors considered to be in operation. Thus, DO level in large scale reactors can be maintained at lower agitation speeds and the power requirement for strain No. 4 for high-strength ammonium treatment will be almost equivalent to that for low-strength ammonium treatment in conventional aerobic nitrification process.

3.3. Wastewater from a chemical company [23]

3.3.1. Introduction

Some wastewaters from chemical companies or power-generation plants contain a high concentration of ammonium and a small amount of BOD. In this section, No. 4 was applied to a wastewater from a chemical company to assess the possibility of the efficient biological treatment of high-strength ammonium.

3.3.2. *Materials and methods*

3.3.2.1. *Wastewater used*

The wastewater (WC) was supplied by a Japanese chemical company. The main characteristics of the WC are as follows: pH 10.6, total COD concentration of 2280 mg/l, total BOD concentration of less than 2 mg/l, total-nitrogen concentration of 4840 mg/l, and ammonium-nitrogen concentration of 4800 mg/l. In each experiment, the pH of the original WC was adjusted to approximately 7.5 by 5N H_2SO_4, and the ammonium concentrations of pH-adjusted WC was diluted to approximately 1000 mg/l unless specifically described.

The experimental procedures in this section were similar to those in Section 3.2.

3.3.3. *Results*

3.3.3.1. *Ammonium removal in the repeated-batch experiment*

Figure 7 shows the change in the ammonium concentration over times in a repeated-batch experiment at 30°C, and **Figure 8** shows the change in the number of No. 4 cells during the same experiment. More than 90% of ammonium was removed within 24–30 h, and the number of No. 4 cells varied between 10^8 and 10^{10} cells/ml. The average ammonium removal rate during the experimental period was 1.1 kg-N/m^3/day. Between 620 and 800 h, the operation was stopped, and the jar fermenter was maintained static at room temperature (average 10°C). When the operation was resumed, ammonium removal was observed without any delay, indicating that the interruption in the operation exerted no adverse effect on the activity of No. 4. In these experiments, the pH values were fluctuated between 7 and 8, which are within the optimal pH range of No. 4. Total amounts of nitrite, nitrate, and exhausted ammonium from the reactors were less than 2% of inlet nitrogen, and thus the majority of inlet ammonium was converted into N_2 gas and the cellular nitrogenous compounds.

Figure 7. Ammonium concentration in the wastewater from a chemical company during repeated-batch treatment with No. 4 at 30°C [23].

3.3.3.2. Ammonium removal at initial ammonium concentrations of 1000, 2000, and 5000 mg NH_4^+-N/l

Figure 9 shows the ammonium removal obtained with initial ammonium concentrations of approximately 5000 mg NH_4^+-N/l, 2000 mg NH_4^+-N/l and 1000 mg NH_4^+-N/l. For concentrations of 5000 mg NH_4^+-N/l and 2000 mg NH_4^+-N/l, an intermittent supply of 20 g of trisodium citrate dihydrate was introduced, as indicated by the arrows in **Figure 9**. The average ammonium removal rates for 1000, 2000, and 5000 mg NH_4^+-N/l were 0.63, 0.96, and 0.92 kg-N/m^3/day, respectively. This indicates that even ammonium concentrations higher than 1000 mg NH_4^+-N/l were removed efficiently by supplying a sufficient amount of the carbon source.

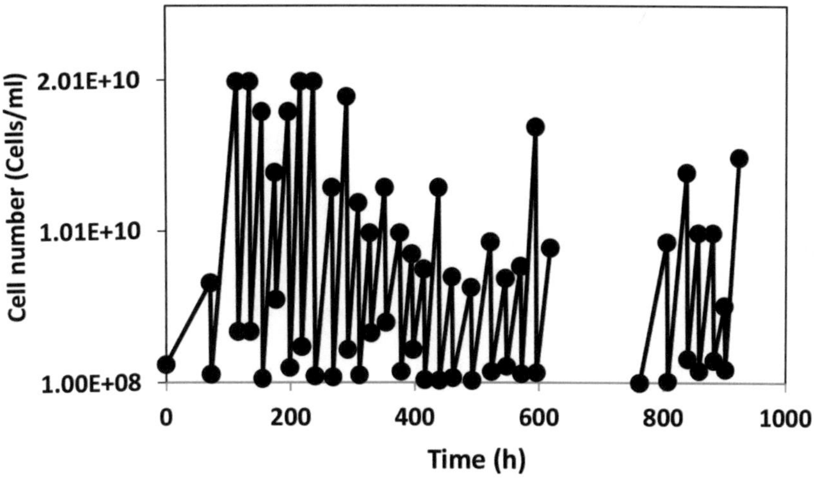

Figure 8. Change in the number of No. 4 cells in the same experiment shown in **Figure 7** [23].

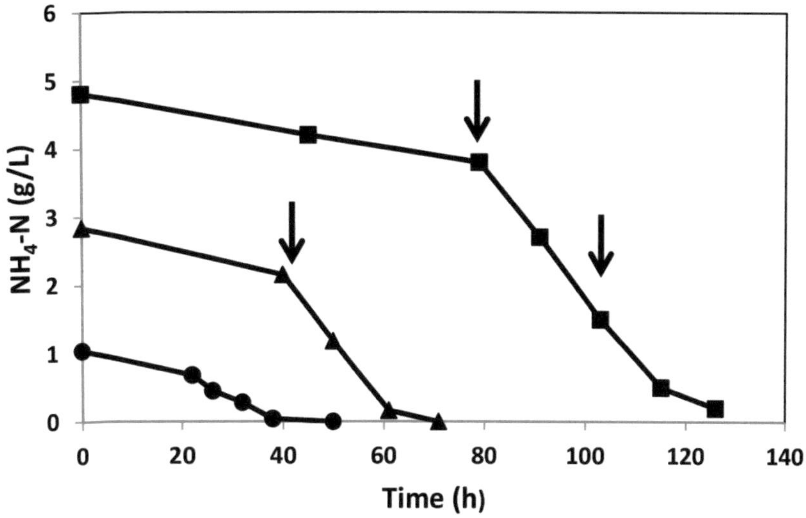

Figure 9. Change in the initial ammonium concentrations when 5000 mg-NH_4^+-N/l (■), 2000 mg-NH_4^+-N/l (▲), and 1000 mg-NH_4^+-N/l (●) of wastewater from a chemical company were used in a batch culture. The arrows indicate the times at which citrate was added [23].

3.3.3.3. Ammonium removal under high salt conditions

NaCl was then added to the WC to a final concentration of 3%, and repeated-batch treatment was conducted using protocol similar to that described in Section 3.2. The result was similar to that in Section 3.2. The ammonium removal rate reached 1.0 kg-N/m^3/day with the four-batch operation after the gradual acclimation of No. 4 to the saline medium.

3.4. Coking wastewater [24]

3.4.1. Introduction

Coking wastewater (CW), which originates from the process of destructive distillation of coal at high temperatures in the absence of air, has been a severe problem. Phenols are the major constituents of the coking wastewater and seriously inhibit various biological reactions, especially the nitrification reaction. Conventional biological treatment for CW is difficult mainly due to refractory substances. When high-strength ammonium is involved in CW, BOD in the wastewater is not sufficient to complete the removal of ammonium. In this section, first, phenol-degradation ability by No. 4 was confirmed, and No. 4 was applied to a coking wastewater supplied by a chemical company to assess the effects of biological treatment of high-strength ammonium and phenol using a 1-l jar fermenter.

3.4.2. Materials and methods

Medium: A synthetic medium described above was used in a preculture, using lactate as a carbon source.

3.4.2.1. Wastewater used

The coking wastewater (CW) was supplied by a Japanese chemical company. The primary characteristics of the CW are as follows: pH 8.5, total COD concentration of 5200 mg/l, ammonium-nitrogen concentration of 800 mg/l, and phenol concentration of 820 mg/l. In each experiment, the pH of the original CW was adjusted to approximately 7.5 by 5N H_2SO_4, and the pH-adjusted CW was diluted arbitrarily.

3.4.2.2. Experimental procedure

The synthetic medium was prepared in the preculture of No. 4 containing phenol (0.2 g/l) only as a carbon source and No. 4 was precultivated for 3 days and the preculture was centrifuged at 10,000 rpm for 10 min. The collected cells of No. 4 were washed with 0.1 M phosphate buffer two times and the cells were inoculated into the synthetic medium, which was devoid of lactate and contained only phenol as a carbon source and utilization of phenol by No. 4 was tested.

In CW treatment, precultured No. 4 cells were introduced into different dilution CW and the growth of No. 4 was confirmed at 50% dilution in shaking flasks. The diluted CW was added with lactate and No. 4 culture in a jar fermenter and ammonium removal test was conducted.

3.4.2.3. *Analytical method*

For phenol concentration determination, the chemical analysis kit for phenol (LR-PNL, Kyoritsu Chemical-Check Lab., Corp., Tokyo, Japan) was used. The initial and final values of TOC in the prepared solution were determined. The crude coking wastewater was streaked on the LB medium and the synthetic agar medium and no colonies appeared after 3 days of incubation. Therefore, indigenous microorganisms in the crude coke-production wastewater were negligible in number.

The air was supplied to the CW sample containing lactate in the jar fermenter for 4 days and neither the removal of lactate nor ammonium was observed, and thus the air-borne microorganisms were neglected.

3.4.3. *Results*

3.4.3.1. *Availability of phenol by No. 4*

Complete removal of ammonium and phenol in the synthetic medium were confirmed when phenol was added as a sole carbon source and the growth of No. 4 (data not shown). The ammonium removal rates using phenol as a carbon source were 0.098–0.12 kg-N/m^3/day, which is approximately one-tenth of the rate when organic acids were used as a carbon source [20]. However, these data were approximately 10-fold higher than the rate in conventional nitrification-denitrification method. When the initial phenol concentration was 600–700 mg/l, this includes 459–535 mg/l of carbon. If the C/N ratio of cell synthesis was 10, consumption of 600–700 mg/l of phenol corresponded with the consumption of only approximately 50–60 mg/l of ammonium-nitrogen. This suggests that for complete removal of high-strength of ammonium in CW, addition of available carbon for No. 4 is needed.

The No. 4 culture was directly mixed with crude-coking wastewater with fortified lactate in a jar fermenter, but removal rates of ammonium and lactate were significantly decreased, presumably toxic substances in coking wastewater inhibited the activity of No. 4. When CW was diluted, the normal growth of No. 4 was observed at 50% dilution. Then, 50% of dilution CW wastewater was mixed with No. 4 preculture and 4 g/l of lactate. The result is shown in **Figure 10**. The initial ammonium-nitrogen concentration, phenol concentration, and lactate concentration were 420 mg/l, 380 mg/l, and 4 g/l, respectively. The ammonium removal rate was 1.8 kg-N/m^3/day and phenol removal rate was 0.7 kg/m^3/day. Phenol removal rate 0.7 kg/m^3/day was two times larger than that in the previous report [25].

COD in the initial wastewater containing lactate was 12,000 mg/l, and after the treatment, this value decreased to 2830 mg/l. The COD of 50% diluted CW was 2130 mg/l. Thus, this ammonium treatment was primarily undertaken by No. 4 by consumption of added lactate and indigenous phenol. As coking wastewater contained some other carbon substances not available for No. 4, further treatment may be needed for complete treatment of COD after this system.

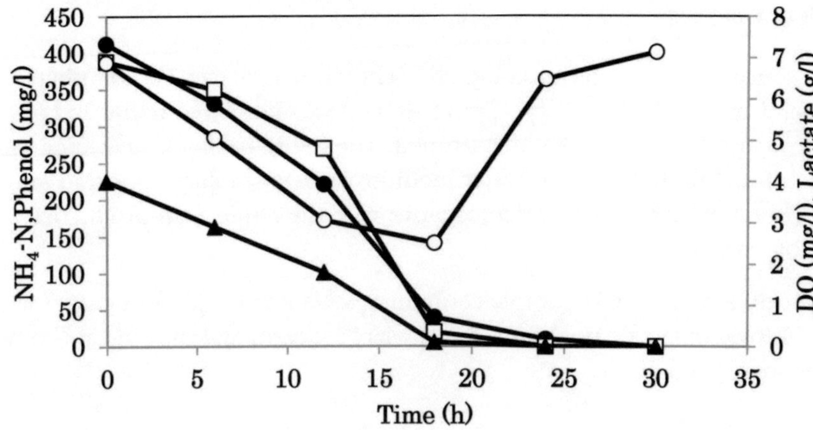

Figure 10. Removal of ammonium, phenol, and lactate in coking wastewater by No. 4. Symbols: NH_4^+-N concentration (•), phenol concentration (□), lactate concentration (▲), and dissolved oxygen (DO) concentration (○) [24].

3.5. Preparation of organic acid solution for No. 4 [26]

3.5.1. Introduction

As No. 4 utilizes primarily organic acids as a carbon source and no sugar is available, chemical agents of citrate or lactate were used as a carbon source in the previous sections. In practical treatment, inexpensive production and supply of organic acids is a key for the materialization of No. 4 in ammonium treatment. In this section, anaerobic fermentation was conducted and then a mixture of high organic acid solution was prepared and this mixture was supplemented with two high-ammonium and low-carbon wastewaters by balancing C/N ratio around to 10 and the effectiveness of the prepared organic acid solution was confirmed.

3.5.2. Materials and methods

3.5.2.1. Wastewaters

The leachate wastewater from a landfill area in B city where the city garbage was land filled was used for ammonium treatment. The total organic carbon (TOC) and ammonium concentration were 4310 mg/l and 880 mg NH_4^+-N/l, respectively.

For a sample containing high NH_4^+-N concentration and the least amount of carbon, anaerobically digested sludge wastewater used in Section 3.2 was used. This contained approximately 900 mg NH_4^+-N/l and almost no BOD was used for ammonium treatment.

3.5.2.2. Preparation of the organic acid solution

Forty milliliter leachate wastewater and 20 g of glucose were mixed in a 1-l plastic container and statically incubate at 30°C for 2 weeks. The TOC and concentrations of eight organic acids in the prepared solution were determined. The volume of organic acid solution was determined so as to be C/N 10.

Experiment 1: Ammonium treatment of the leachate wastewater using No. 4 culture and organic acid solution 230 ml of leachate wastewater, 30 ml of No. 4 culture, and 40 ml of organic acid solution were mixed and the ammonium treatment was conducted in a jar fermenter.

Experiment 2: Ammonium treatment of anaerobically digested sludge using No. 4 culture and organic acid solution 180 ml of the wastewater, 30 ml of No. 4 culture, and 90 ml of organic acid solution were mixed and the ammonium treatment was carried out.

3.5.3. Results

3.5.3.1. Prepared highly concentrated organic acid solution

After 2- week anaerobic incubation of the leachate wastewater, the resulting organic acid solution contained 20,049 mg/l of TOC and 52,103 mg/l of total organic acid content of eight types, as shown in **Table 5**. The estimated carbon content from the organic acid data was 20,754 mg/l, as shown in **Table 5**. As the carbon contents in the TOC and organic acid solution were almost similar, TOC was used as an indicator to adjust to the necessary carbon content required to treat ammonium completely by balancing C/N ratio 10.

3.5.3.2. Experiment 1

The result is shown in **Figure 11**. The initial TOC was 7017 mg/l and the initial NH_4^+-N concentration was 659 mg/l. The initial value of eight kinds of organic acids was 17,750 mg/l in which the estimated carbon content was 6500 mg/l (**Table 6**). The final value of TOC was 900 mg/l and the final carbon value of eight kinds of organic acids was 817 mg/l, as show in **Table 6**. Complete ammonium removal was observed and thus the effectiveness of the use of organic acid solution and the use of TOC as an index to determine C/N ratio was confirmed. The ammonium removal rate was 1.1 kg-N/m^3/day.

	Content (mg/l)	Carbon content (mg/l)
Oxalate	250	64
Citrate	250	94
Lactate	41202	16481
Formate	479	125
Acetate	4750	1900
Propionate	2402	613
iso-Butyrate	250	102
n-Butyrate	2520	1375
Total	52,103	20,754

Table 5. Organic acid distribution and carbon content in the prepared organic acid solution [26].

3.5.3.3. Experiment 2

In Section 3.2, the high-strength ammonium from anaerobically digested sludge was removed using No. 4 with addition of citrate. A similar sample that contained 900 mg NH_4^+-N/l and 20 mg/l of organic content indicated that the available carbon for No. 4 is scarce, and supplementation of the organic acid solution is essential for complete removal of ammonium. For the initial 180 ml sludge sample, 90 ml of organic acid solution and 30 ml of No. 4 culture were mixed, and the ammonium removal was measured. The results are shown in **Figure 12** and **Table 7**. Similarly, the initial NH_4^+-N concentration 635 mg/l was completely removed and the ammonium removal rate was 0.8 kg N/m³/day.

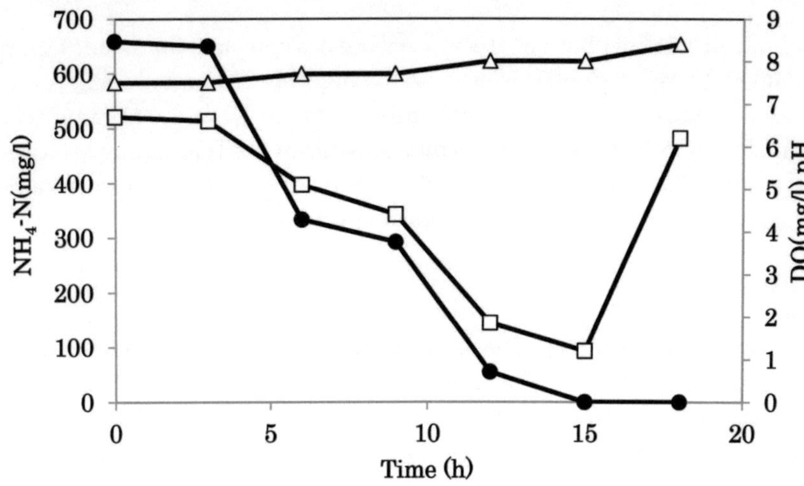

Figure 11. Ammonium removal when leachate wastewater was treated with No. 4 culture and organic acid solution. Symbols: NH_4^+-N (•), dissolved oxygen concentration (DO) (□), and pH (△) [26].

	Initial carbon content (mg/l)	Final carbon content (mg/l)
Oxalate	64	64
Citrate	94	94
Lactate	3600	100
Formate	65	65
Acetate	1680	100
Propionate	842	156
iso-Butyrate	102	102
n-Butyrate	136	136
Total	6500	817

Table 6. Change in the initial and final carbon contents of organic acids in **Figure 11** [26].

	Initial carbon content (mg/l)	Final carbon content (mg/l)
Oxalate	6.4	6.4
Citrate	9.4	9.4
Lactate	6000	100
Formate	22	6.5
Acetate	720	138
Propionate	181	64
iso-Butyrate	10.2	10.2
n-Butyrate	365	13.6
Total	7314	348

Table 7. Change in the initial and final carbon contents of organic acids in **Figure 12** [26].

Factors	No. 4 method	Conventional method	Anammox method
Number of reactor	1	2	2
Reactor cost	Low	High	High
Addition of O_2	Essential	Essential	Essential
Organic matter	Essential	Essential	No use
Maintenance of microorganisms	Easy	Difficult	Difficult
Activity persistence	Long	Short	Short
Application to high C/N waste	Applicable	Pre-treatment essential	Pre-treatment essential
N_2 production speed	High	Low	High
System control	Easy	Difficult	Difficult

Table 8. Comparison of No. 4, anammox, and conventional methods.

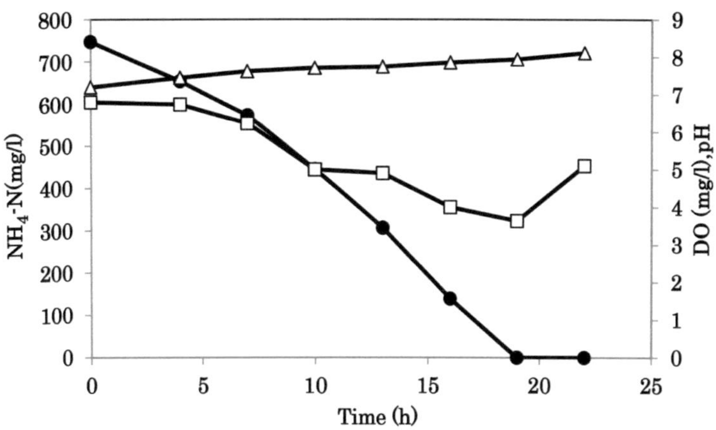

Figure 12. Ammonium removal when anaerobically digested sludge was treated with No. 4 culture and organic acid solution. Symbols: NH_4^+-N (•), dissolved oxygen concentration (DO) (□),and pH (△) [26].

4. Additional characteristics of No. 4

In wastewater treatment, the excess sludge is inevitably produced, and new treatment possibilities of excess sludge containing No. 4 are presented in the following sections.

4.1. Suppression of growth of plant pathogens by No. 4 [19]

4.1.1. Introduction

In this section, it was presented that the *A. faecalis* No. 4 (No. 4) exhibited a suppressive effect on the damping-off caused by the plant pathogen *Rhizoctonia solani* on soil [27].

4.1.2. Materials and methods

4.1.2.1. Preparation of culture broth and cell suspension

Fifty milliliters of culture broth of No. 4 after cultivation in L medium was mixed with 150 g of soil in a pot. For the treatment consisting of only the cell suspension, the cells of No. 4 were collected by centrifugation, and the sedimented cells were suspended in sterile distilled water and 50 ml of the cell suspension was mixed with 150 g of soil.

4.1.2.2. Soil treatments and inoculation of soil with R. solani

These procedures were described in detail in a previous section [19].

4.1.2.3. Plant growth

For each treatment, three pots were prepared. Tomato (*Lycopersicum esculentum*) seeds were germinated on 2% agar plates at 30°C for 2 days, and nine germinated seeds were planted in each pot and incubated in a grown chamber.

4.1.3. Results

The result of the effect of No. 4 culture broth on the damping-off of tomato seedlings caused by *R. solani* in soils is shown in **Figure 13**. In a pot that was not infested with *R. solani*, all the seedlings grew normally and no disease was apparent. In a pot infested with only *R. solani*, the percentage of diseased plants was 78–82% and the shoot weight and leaf length were markedly decreased in soil. However when the culture broth of No. 4 was introduced into the soil, the percentage of diseased plants was reduced to 17%. When the cell suspension of No. 4 was applied to soil, the percentage of diseased plants in soil was similar to that soil treated with No. 4 culture broth (data not shown).

The finding suggests that the treatment with No. 4 cells is effective for plant disease control.

Figure 13. *A. feacalis* No. 4 exhibits suppressive effect on plant pathogens. Pot ③: Tomato seedlings with plant pathogens. Pot ①: Tomato seedling without plant pathogens. Pot ②: *A. feacalis* No. 4 was introduced to pot ③.

4.2. Reduction of methane production from rumen of cows [28]

4.2.1. *Introduction*

Enteric methane (CH_4) production from livestock is a significant source of greenhouse gas. It is reported that nitrate (NO_3) suppresses enteric CH_4 production. However, the reduction of NO_3 to nitrite (NO_2) in the rumen results in the accumulation in NO_2, which is toxic to livestock.

A denitrifying bacterium, *A. faecalis* No. 4, was coincubated with a low concentration of NO_3 (2 mmol/l) and the *in vitro* CH_4 production was tested.

4.2.2. *Materials and methods*

4.2.2.1. *Rumen liquid*

The rumen liquid which was collected from Holstein cows and No. 4 cells obtained after centrifugation of culture broth prepared in synthetic medium were mixed with 2 mmol/l NO_3. The mixture was placed in a 1 l jar fermenter and CH_4 production was monitored under anaerobic condition.

4.2.3. *Result*

The methane production from rumens is shown in **Figure 14**. When No. 4 and 2 mmol/l NO_3 were mixed, methane production showed a significant decrease without causing an adverse impact on anaerobic fermentation in rumens. This suggests a possibility of re-use of No. 4, which was produced as excess sludge after treatment of high-strength ammonium.

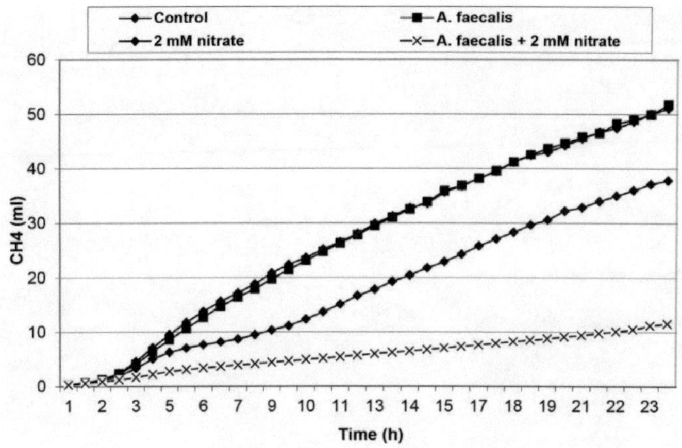

Figure 14. Methane production from rumens when No. 4 was introduced [28].

5. Conclusions

Table 8 shows comparison among conventional ammonium treatment method, anammox method and No. 4 method. No. 4 has many advantages over other methods. No. 4 was effective to remove high-strength of ammonium in several wastewaters when organic acids are supplied. The excess cells of No. 4 are produced during treatment of ammonium because the cell growth of No. 4 is associated with ammonium removal. Possibility of the re-use of the excess cells in agricultural areas was presented. How to collect organic acids as carbon source for No. 4 is a problem to be solved. Production of high concentration of organic acids in anaerobic fermentation and the use of the produced organic acid solution to wastewaters were shown as one possible method.

Acknowledgements

These researches were supported by many organizations of Japan; Able Cooperation (Tokyo), Kanagawa Prefectural Livestock Industry Research Institute, Yokohama Municipal Sewage Treatment Center, Several Chemical Companies, Obihiro University of Agriculture and Veterinary Medicine and Kanagawa Prefectural Agricultural Research Institute.

Author details

Makoto Shoda

Address all correspondence to: mshoda@res.titech.ac.jp

Chemical Resources Laboratory, Tokyo Institute of Technology, Yokohama, Japan

References

[1] Neerackal G M, Ndegwa P M, Joo H S, Wang X, Frear C S, Harrison J H, Beutel M W. Potential application of *Alcaligenes faecalis* strain No. 4 in mitigating ammonia emissions from dairy wastewater. Bioresour. Technol. 2016;206:36-42.

[2] Van Dongen U, Jetten MSM, van Loosdrecht MCM. The SHARON-Anammox process for treatment of ammonium rich wastewater. Water Sci. Technol. 2001;44:153-160.

[3] Van de Graaf AA, deBruijin P, Robertson LA, Jetten MSM, Kuenen JG. Autotrophic growth of anaerobic ammonium-oxidizing micro-organisms in a fluidized bed reactor. Microbiology. 1996;142:2187-2196.

[4] Furukawa K, Inatomi Y, Qiao S, Quan L, Yamamoto T, Isaka K, Sumio T. Innovative treatment system for digester liquor using anammox process. Bioresour. Technol. 2009;100:5437-5443.

[5] Chen PZ, Li J, Li QX, Wang YC, Li SP, Ren TZ, Wang LG. Simultaneous heterotrophic nitrification and aerobic denitrification by bacterium *Rhodococcus* sp. CPZ24. Bioresour. Technol. 2012;116:266-270.

[6] Chen Q, Ni JR. Ammonium removal by *Agrobacterium* sp. LAD9 capable of heterotrophic nitrification-aerobic denitrification. J. Biosci. Bioeng. 2012;113:619-623.

[7] Khardenavis AA, Kapley A, Purohit HJ. Simultaneous nitrification and denitrification by diverse *Diaphorobacter* sp. Appl. Microbiol. Biotechnol. 2007;77:403-409.

[8] Yang XP, Wang SM, Zhang DW, Zhou LX. Isolation and nitrogen removal characteristics of an aerobic heterotrophic nitrifying-denitrifying bacterium, *Bacillus subtilis* Al. Bioresour. Technol. 2011;102:854-862.

[9] Yao S, Ni JR, Chen Q, Borthwick AGL. Enrichment and characterization of a bacteria consortium capable of heterotrophic nitrification and aerobic denitrification at low temperature. Bioresour. Technol. 2013;127:151-157.

[10] Zhang J, Wu P, Hao B, Yu Z. Heterotrophic nitrification and aerobic denitrification by the bacterium *Pseudomonas stutzeri* YZN-001. Bioresour. Technol. 2011;102:9866-9869.

[11] Zhang QL, Liu Y, Ai GM, Miao LL, Zheng HY, Liu ZP. The characteristics of a novel heterotrophic nitrification-aerobic denitrification bacterium, *Bacillus methylotrophicus* strain L7. Bioresour. Technol. 2012;108:35-44.

[12] Zhao B, He YL, Hughes J, Zhang XF. Heterotrophic nitrogen removal by a newly isolated *Acinetobacter calcoaceticus* HNR. Bioresour. Technol. 2010;101:5194-5200.

[13] Zhu L, Ding W, Feng LJ, Kong Y, Xu J, Xu XY. Isolation of aerobic denitrifiers and characterization for their potential application in the bioremediation of oligotrophic ecosystem. Bioresour. Technol. 2012;108: 1-7.

[14] Su JJ, Yeh KS, Tseng PW. A strain of *Pseudomonas* sp. isolated from piggery wastewater treatment systems with heterotrophic nitrification capability in Taiwan. Curr. Microbial. 2006;53:77-81.

[15] Liu Y, Hu T,Song Y, Chen H, Lv Y. Heterotrophic nitrogen removal by *Acinetobacter* sp. Y1 isolated from coke plant wastewater. J. Biosci. Bioeng. 2015;120: 549-554.

[16] Ren YX, Yang L, Liang X. The characteristics of a novel heterotrophic nitrifying and aerobic denitrifying bacterium, *Acinetobacter junii* YB. Bioresour. Technol.2014;171:1-9.

[17] Zheng HY, Liu Y, Gao XY, Ai GM, Liu ZP. Characterization of a marine origin aerobic nitrifying-denitrifying bacterium. J. Biosci. Bioeng. 2012;114:33-37.

[18] Joo HS, Hirai M, Shoda M. Nitrification and denitrification in high-strength ammonium by *Alcaligenes faecalis*. Biotechnol. Lett. 2005; 27:773-778.

[19] Honda N, Hirai M, Ano T, Shoda M. Control of tomato damping-off caused by *Rhizoctonia solani* by the heterotrophic nitrifier *Alcaligenes faecalis* and its product, hydroxylamine. Ann. Phytopathol. Soc. Japan. 1999;65:153-162.

[20] Shoda M, Ishikawa Y. Heterotrophic nitrification and aerobic denitrification of high-strength ammonium in anaerobically digested sludge by *Alcaligenes faecalis* No. 4. J. Biosci. Bioeng. 2014;117:737-741.

[21] Joo HS, Hirai M, Shoda M. Piggery wastewater treatment using *Alcaligenes faecalis* strain No. 4 with heterotrophic nitrification and aerobic denitrification. Water Res. 2006;40: 3029-3036.

[22] Obaga D, Mace S, Mata-Alvarez J. Biological nutrient removal by a sequencing batch reactor (SBR) using an internal organic source in digested piggery wastewater. Bioresour. Technol. 2005; 96:7-14.

[23] Shoda M, Ishikawa Y. Heterotrophic nitrification and aerobic denitrification of a waste-water from a chemical company by *Alcaligenes faecalis* No. 4.Int. J. Water Wastewater Treat. 2015;1(2) http://dx.doi.org/10.16966/2381-5299.111.

[24] Shoda M, Ishikawa Y. Removal of High-Strength of Ammonium and Phenol from Coking Wastewater by *Alcaligenes faecalis* No. 4. J. Appl. Biotechnol. Bioeng. 2006;1(3):00014. doi:10.15406/jabb.2016.01.00014.

[25] Ge Q, Yue X, Wang G. Simultaneous heterotrophic nitrification and aerobic denitrification at high initial phenol concentration by isolated bacterium *Diaphorobacter* sp. PD-7. Chin. J. Chem. Eng. 2015; 23:835-841.

[26] Shoda M, Ishikawa Y. Utilization of organic acid solution prepared for high-strength ammonium treatment using *Alcaligenes faecalis* No. 4. Int. J. Water Wastewater Treat. 2016;2(5):doi http://dx.doi.org/10.16966/2381-5299.131.

[27] Cook RJ. Twenty-five years of progress towards. In Biological Control of Soil-Borne Plant Pathogens (Hernby D. ed.), CAB International, Wallingford. 1990 p.1-14.

[28] O'Brien M, Shoda M, Nishida T, Takahashi J. Synergistic effect of *Alcaligenes faecalis* and nitrate to reduce *in vitro* rumen methanogenesis, Proceedings of the 5th International Conference on Greenhouse Gases and Animal Agriculture, June 23-26, Dublin, Ireland, 2013.

Chapter 4

Biodegradation of Ammonia in Biofiltration Systems: Changes of Metabolic Products and Microbial Communities

Yiu Fai Tsang, Ya-nan Wang, Huawei Wang,
Yi Yang, Yuanhui Zhang and Hong Chua

Additional information is available at the end of the chapter

http://dx.doi.org/10.5772/intechopen.68155

Abstract

In the first stage, the feasibility of using the waste materials from coal power plants (i.e., coal slag) and landscapes (i.e., wood chip and compost) as packing media in various biofiltration systems for ammonia (NH_3) removal was investigated. In the second stage, the optimized biotrickling system packed with coal slag was employed to investigate the effects of inlet concentration on NH_3 treatment performance. A complete NH_3 removal was achieved at concentrations of up to 250 ppm at an empty bed retention time of as low as 8 s, which is shorter than most previously reported biofiltration systems. Results of metabolic product analysis indicated that half of introduced NH_3 was oxidized to nitrate and the rest was converted to ammonium ion at low loadings, while nitrite and ammonium ions predominate at high loadings. A bacterial community shift was observed with regard to the loading rates and pH conditions. In addition, there were no common operating problems, such as clogging and compaction, in the operation for more than 1 year.

Keywords: biofilter, biotrickling filter, ammonia removal, nitrogen mass balance, microbial community

1. Introduction

Ammonia is characterized as a colorless, toxic, reactive, and corrosive gaseous pollutant with a strong and repulsive smell [1]. NH_3 is emitted as a by-product of different industrial processes, such as wastewater treatment, composting, livestock production, and petrochemical refining [2, 3]. Its emission causes significant odor nuisance, health impacts, and environmental problems. It has been reported that exposure to NH_3 above 1 ppm could cause nausea, headaches, bronchial tract irritation, and burning sensation in the eyes and skin [4, 5]. It is crucial to control NH_3 emission to protect public health and the environment.

The conventional ammonia treatment methods are based on physical and chemical processes, such as adsorption, scrubbing, and chemical oxidation. Unfortunately, these are expensive and produce secondary waste that may require further treatment or disposal, thereby creating additional environmental problems [6, 7]. Biofiltration is an emerging technology for the control of odor and ammonia from contaminated air streams [8–10]. Studies suggest that compost-based biofilters are the most cost-effective for low-concentration ammonia removal in the agricultural industry due to their low operating and maintenance costs. Other supplementary materials (e.g., wood chips) are commonly added to reduce pressure drop and provide a solid-phase buffer [8]. The primary mechanism of biofiltration is the heterogeneous biochemical process controlled by either mass transfer or biochemical reaction or both. Pollutants are transferred from the air to the water layer or to the biofilm attached on the packing media by adsorption or absorption. The sorbed contaminants in the biofilm are degraded by microorganisms into carbon dioxide, water, biomass, and energy [6, 7].

The main functions of the packing media are to provide contact between the gaseous contaminants and the active biofilm and to distribute water and nutrients on the packing surface [8, 11]. Biofilter performance and operating cost are affected by the media characteristics, such as surface area, mechanical properties, buffer capacity, nutrient availability, porosity, and water retention capacity, hence providing an ideal environment for microbial growth [10, 11]. Therefore, the selection of suitable supporting materials and operating mode is an important aspect of a successful biofiltration process.

There are three general classifications of media, namely, natural, inert, and synthetic. Natural materials, including peat, soil, and compost, are generally chosen as biofilter media because they are inexpensive and have a wide diversity of indigenous microorganisms [11, 12]. In addition, several research studies have revealed that natural packing materials provide superior performance in ammonia treatment [8, 13]. Nonetheless, natural-based biofilters are often plagued by common operating problems, such as compaction and decomposition, hence resulting in high pressure drop and air channeling. Common inert materials used in biofiltration include glass beads, perlite, and porous ceramics. Inert materials are difficult to compact. Moreover, they maintain a stable composition during long-term operation. Consequently, they could be used as an alternative to natural media [11]. However, their wide application is stifled due to high material costs and nutrient deficiency.

Different natural and inert packing materials have been successfully applied in biofiltration systems [10, 14, 15]. Likewise, extensive studies have focused on the selection of filter materials and on the optimization of reactor design and operating criteria to obtain efficient ammonia removal in biofiltration systems. However, it is difficult to evaluate the efficiency of various filter materials, because the simultaneous comparison of natural and inert packing media has not been clearly determined under the same conditions. In addition, there are only limited studies on inert packing materials and trickling operations. In our previous study, an attached growth bioreactor packed with coal slag was successfully utilized for domestic wastewater treatment both in lab- and pilot-scale experiments, indicating that coal slag is a viable supporting material for biofilm attachment and long-term operation [16–18]. The relatively high adsorption capacity of coal slag is also an advantage in the biotrickling filter.

The aim of this study is to determine the feasibility of using recycled wastes as packing media in biofiltration systems for ammonia removal. Potential packing materials were characterized and selected for further investigation, and different operating modes of reactors, namely, biofilter and biotrickling reactor, were also evaluated in terms of ammonia elimination capacities. The removal mechanisms and the inhibitory effects were also investigated through the mass balance analysis.

2. Materials and methods

2.1. Experimental setup

The biofiltration system consisted of four conventional biofilters and one biotrickling reactor (**Figure 1a** and **b**), which were packed with selected waste materials (**Table 1**), and could be operated independently to one another or as replicates. Reactors 1–5 were constructed with cylindrical PVC plastic buckets with a sealable lid. Each reactor has a effective volume of 14.8 l (D = 305 mm; effective height = 204 mm). Perforated plastic plates were installed as support for the packing materials and to provide a good gas and liquid distribution. The biotrickling filter (Reactor 5) consisted of a recirculation system (10 l nutrient tank + peristaltic pump (Masterflex, 7518–10) with an anti-clogging sprinkler), a pH controller (model 8156, Orion pH electrode), and a NaOH feeding tank. The recirculation liquid was regularly replaced after each 7-day operation, in order to prevent accumulation of toxic by-products in the reactor. The flow rate of odorous gas stream entering each reactor was controlled by five adjustable panel mounted rotameters (CZ-3246-24, Cole-Parmer) and a mass flow controller. The sampling system monitored the inlet and outlet ammonia concentrations, ambient temperature, and temperature of each reactor and regulated the mass flow controller to provide steady and desirable ammonia concentrations from a pure NH_3 gas cylinder (S.J. Smith Co., USA). Throughout all the experimental runs, the reactors were operated at room temperature of about 25°C. In Reactors 1–3, biofilm development was employed by natural selection without an external inoculum. In Reactor 4, diluted activated sludge solution was completely mixed with wood chips and coal slag as a seed. For Reactor 5 (biotrickling filter), the acclimated microbial broth was added into the fresh mineral medium without $(NH_4)_2SO_4$. The solution was sprayed on the top of the filter bed with a flow rate of 100 ml min^{-1}. Thereafter, the synthetic odorous gas with a relatively low NH_3 concentration (10–20 ppm) was introduced into the reactor for the start-up process.

2.2. Waste materials

Six waste materials, including shredded hardwood mulch, wood mulch fine compost, chipped hardwood mulch, mushroom compost, landscape wood chips, and coal slag, were selected to determine the feasibility of using these materials as packing media for ammonia removal. The physicochemical characteristics of the waste materials were analyzed. In the continuous

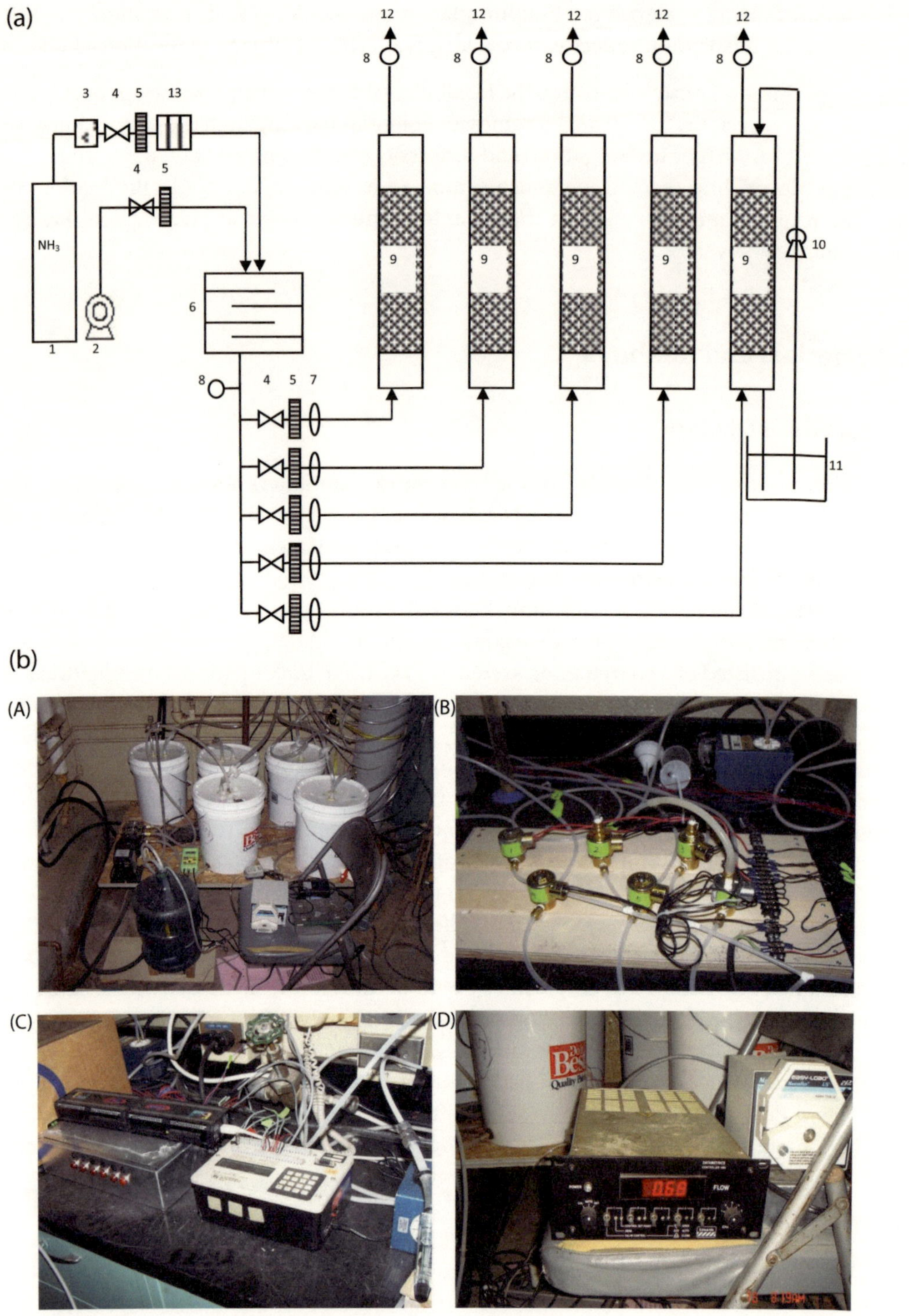

Figure 1. (a) Schematic diagram of the biofiltration system: (1) gas cylinder, (2) compressed air, (3) pressure regulator, (4) valve, (5) flowmeter, (6) gas mixing chamber, (7) air filter, (8) sampling points, (9) filter medium, (10) peristaltic pump, (11) recirculation tank, (12) gas outlet, and (13) mass flow controller. (b) Photos of the biofiltration system: (A) reactors, (B) sampling systems, (C) datalog/control system, and (D) mass flow controller.

	Supporting materials	Operating mode	Inoculation	Feeding solution	Spraying frequency
Reactor 1	Wood chips (100%)	Conventional	No	Water	20 min d^{-1}
Reactor 2	Wood chips (50%) + compost (50%)	Conventional	No	Water	20 min d^{-1}
Reactor 3	Wood chips (50%) + coal slag (50%)	Conventional	No	Water	20 min d^{-1}
Reactor 4	Wood chips (50%) + coal slag (50%)	Conventional	Mixed with sludge	Water	20 min d^{-1}
Reactor 5	Coal slag (100%)	Trickling	Inoculated with sludge	Mineral nutrient	Continuous spraying

Table 1. Operating conditions used in the preliminary study for ammonia removal.

experiment for ammonia removal, three potential materials were selected from six wastes as packing media based on their desirable properties, namely, landscape wood chips, mushroom compost, and coal slag.

2.3. Microorganisms and inocula

The sludge was obtained from the return activated sludge stream at a local wastewater treatment plant in Urbana, Illinois, USA. The autotrophic ammonia-oxidizing bacteria (AOB) used in the study of the biotrickling reactor were prepared by acclimating activated sludge with selective ammonia medium (**Table 2**) and sufficient aeration (CO_2 source) for a week [19, 20]. Subsequently, the selected microbial strains were transferred to fresh media. After 3–4 transfers (about 3 weeks of acclimation), the inoculum was ready for inoculation into the biotrickling filter. During the acclimation and operation periods, the mineral nutrient that had no supply of nitrogen source for bacterial growth was used for the recirculation liquid.

2.4. Analytical methods

The inlet and outlet ammonia concentrations of each reactor were monitored using a chemiluminescence NH_3 analyzer (Model 17C, Thermo Electron Corporation) with a sampling system composed of six solenoid valves (Parker, C3A) and a data log (Personal Daq/56) and control (Campbell Scientific Inc., 21X) system. Prior to the study, the analyzer was calibrated using

Chemicals	Concentration
$(NH_4)_2SO_4$	8 g l^{-1}
KH_2PO_4	2 g l^{-1}
K_2HPO_4	2 g l^{-1}
$MgCl_2{\cdot}6H_2O$	0.4 g l^{-1}
$CaCl_2{\cdot}2H_2O$	0.05 g l^{-1}

Table 2. Composition of ammonia medium.

standard NH_3 gas with a range of zero to 100 ppm. Ammonia, nitrate, and nitrite in water and solid phases were measured using the colorimetric method (Technicon AA II Continuous-flow AutoAnalyzer) according to the standard methods [21].

2.5. Microbial community analysis

The total DNA from inoculated coal slag and recirculation liquid was extracted by UltraClean Soil DNA kits with inclusion of PCR inhibition removal solution (Mo Bio Laboratories, Solana Beach, CA), followed by the manufacturer's protocol. The intact DNA was confirmed on a 1% agarose gel electrophoresis. 16S rRNA genes of DNA extracted from the samples were amplified by the PCR using the eubacterial primers 338f and 1492r (**Table 3**) [22]. The 5′ end of the forward primer was labeled with 6-carboxyfluorescein (FAM) for terminal restriction fragment length polymorphism (T-RFLP) analysis. The PCR reaction mixture contained 1 x PCR buffer, 2 mM $MgCl_2$, 0.2 mM dNTP, 0.2 μM of each forward and reverse primers, 2 μl of DNA template, and 2.5 U of Ex Taq DNA polymerase (TaKaRa Biomedicals, Otsu, Shiga, Japan) in a 50 ml final volume. The PCR was performed in a thermal cycler (PTC-200 DNA Engine, MJ Research Inc., Reno, NV, USA). Meanwhile, the amplification was done with one denaturing step at 94°C for 5 min, followed by 30 cycles of denaturing at 94°C for 1 min, annealing at 55°C for 1 min, and extension at 72°C for 1.5 min with a final extension step at 72°C for 10 min.

Terminal restriction fragment length polymorphism (T-RFLP) based on the 16S rRNA gene for microbial community analysis was applied to the samples collected from the reactor at various conditions and operation periods. The PCR reactions were carried out in triplicate and were pooled. Following the confirmation of a successful PCR reaction by agarose gel electrophoresis, the PCR products were purified with a PCR purification kit (UltraClean PCR Clean-Up Kit, Mo Bio Laboratories Inc., Carlsbad, CA, USA). The purified fluorescently-labeled PCR products were digested with restriction enzymes: (1) MspI (TaKaRa, China) and (2) RsaI (New England Biolabs, USA) for 3 h at 37°C, followed by an enzyme inactivation step at 65°C for 20 min, and (3) TaqI (TaKaRa, China) for 3 h at 65°C followed by an enzyme inactivation step at room temperature for 20 min. The digested samples were treated by ethanol precipitation to remove excess salt, and the entire 10 μl digested product was analyzed using the ABI Prism 3100 system (Applied Biosystems, USA). The fluorescently-labeled terminal restriction fragments (T-RFs) were separated by capillary electrophoresis to determine the number and size of T-RFs and subsequently compared to the mobility of size standard fragments. Meanwhile, the fragment analysis was conducted using the GeneMapper™ Version 3.7 software (Applied Biosystems, USA). In both cases, the peak areas (related directly to peak fluorescence) of each T-RF relative to the total peak areas were used to determine the relative abundance of individual

Primer	Sequence (5′–3′)	Specificity	References
Eub338f	ACTCCTACGGGAGGCAGC	Bacteria 16S rRNA gene	Amann et al., [34]
1492r	TACCTTGTTACGACTT	Bacteria 16S rRNA gene	Lin and Stahl [35]

Table 3. PCR primers.

microorganisms within the community being examined. The T-RFLP profiles were analyzed using the TAP T-RFLP program at the Ribosomal Database Project II (RDP) website (http://rdp8.cme.msu.edu/html/TAP-trflp.html#introduction).

3. Results and discussion

3.1. Physicochemical properties

The common properties of the waste materials, such as density, porosity, moisture content (MC), and C/N ratio, were determined to establish the background information for process design (**Table 4**). In general, the pH of various materials was relatively neutral, with the range of 6.3–7.8; hence, these are suitable conditions for most bacterial activities. Based on their physicochemical properties, three wastes were selected to study the performance of the continuous treatment of ammonia. Landscape wood chip and mushroom compost were selected according to higher yield, higher moisture content, and more favorable pH. Coal slag was selected for comparison purpose and due to high availability of the materials from commonly used power stations.

3.2. Start-up

All reactors started with an inlet loading of about 1.5 g m^{-3} h^{-1}, corresponding to the NH_3 concentration of 20 ppm and the EBRT of 30 s. After 14-day operation, the stable removal efficiencies of each reactor were achieved with a range of 17–80%. It was found that the immobilization period of Reactor 5 (biotrickling filter) was completed within 6-day operation. This is comparatively shorter than other traditional biotrickling filters packed with inert or inorganic materials; a 1- to 2-week (up to 4 weeks) operation was required for the completion of start-up [23, 24]. The biodegradation does not occur immediately because some species that are capable of degrading the contaminants may not be initially present [25]. Therefore, the poor performance in Reactors 1, 3, and 4 was observed even at the low loading rates during the start-up.

	Density (kg m^{-3})	Porosity	pH	MC (%)	C/N
Shredded hardwood mulch (1)	0.16 ± 0.01	0.68 ± 0.01	6.58 ± 0.09	7.1 ± 0.7	87.3 ± 0.4
Wood mulch fine compost (2)	0.53 ± 0.02	0.49 ± 0.01	7.33 ± 0.17	18.3 ± 1.4	12.8 ± 0.5
Chipped hardwood mulch (3)	0.23 ± 0.01	0.63 ± 0.02	7.85 ± 0.09	8.3 ± 1.4	33.2 ± 0.3
Mushroom compost (4)	0.21 ± 0.01	0.48 ± 0.02	7.96 ± 0.11	23.7 ± 1.3	20.4 ± 0.3
Landscape wood chip (5)	0.21 ± 0.01	0.65 ± 0.02	7.06 ± 0.06	14.9 ± 0.7	61.3 ± 0.2
Coal slag (6)	1.43 ± 0.04	0.68 ± 0.02	6.32 ± 0.10	3.5 ± 0.8	62.0 ± 0.4

Table 4. Physicochemical properties of natural materials and coal slag (n = 5).

The removal of NH_3 in the biotrickling reactor occurred immediately, primarily because of the adsorption on the coal slag and the absorption in the water layer. From Days 3 to 10, the pH of the recirculation liquid decreased from 7.7 to 5.4, indicating that the biofilm had begun to develop and the NH_3 was mainly removed by microbial activities. Nitrite, nitrate, and hydrogen ions were produced during the nitrification process. As a result, the removal rate subsequently increased to more than 80% by Day 6. This was an indication that the bacteria had been acclimated to NH_3.

From **Figure 2**, Reactors 1, 3, and 4 had poor treatment performance (17–30% removal). The compost-based biofilter (Reactor 2) had the best treatment performance among the four biofilters. The pH decreased from 9 to 6.64 during the start-up period, indicating that the inherent source of bacteria was important for the biofilter operation mode (**Figure 3**). In Reactors 1, 3, and 4, the pH values fluctuated between 8 and 9.5. Therefore, adsorption and absorption processes were the main removal mechanisms during the 2-week operation, leading to the poor elimination capacities. Furthermore, due to an insufficient amount of indigenous bacteria in the landscape wood chips, it was found that either there was no developed biofilm or it was inactive in these biofilters.

3.3. Continuous operation under stepwise increase of NH_3 concentration

After the 14-day operation, continuous experiments with a stepwise increase of NH_3 concentrations from 20 to 70 ppm at a constant flow rate of 28.3 l min^{-1} (EBRT = 30 s) were carried out for a period of 5 months to evaluate the performance of the reactors packed with different materials. Ammonia in biofilters is partly retained by adsorption onto the packing media and by absorption into the water fraction of the carrier materials [26] and partly achieved through nitrification by the autotrophic ammonia-oxidizing bacteria (AOB) and nitrite-oxidizing bacteria

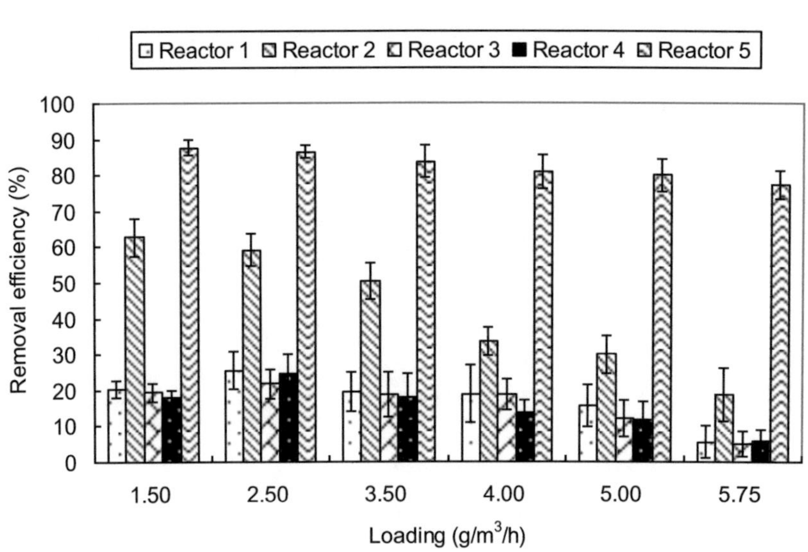

Figure 2. NH_3 removal efficiencies under various loading rates.

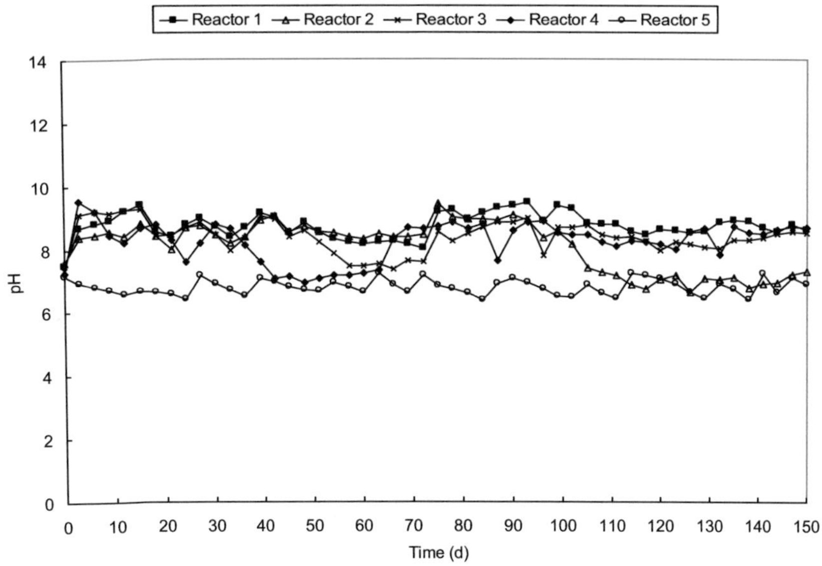

Figure 3. pH profiles during the 150-day continuous treatment.

(NOB) [27]. The results indicated that good bacteria activities with steady and consistent removal (75–88%) were achieved under different loading rates in Reactor 5 only (**Figure 2**). It was found that the pH of Reactor 5 reduced from 7.2 to 6.4 during each replacement of the recirculation liquid, indicating that the ammonia was oxidized by nitrifying bacteria (**Figure 3**). During the change of inlet concentrations, it was observed that various reactors required different adaptation periods to achieve new steady-state conditions. On Day 76, the inlet concentration increased from 40 to 50 ppm, and the removal efficiency significantly decreased from 87 to 70% but was followed by a stable removal efficiency of 80%. The reactor required a longer period to achieve the steady performance at higher NH_3 loading rates.

The mass balance analysis of ammonia oxidation process in Reactor 5 is shown in **Figure 4**. As the loading rate increased up to around 4.5 g m^{-3} h^{-1}, the removal efficiency of Reactor 5 decreased from 87 to 77%. This may be attributed to the nitrifying bacteria activities in the reactor that reached the critical loading rate, whereas the relative abundance of NH_4^+ significantly increased from 28 to 45%. This showed that absorption became the dominant mechanism for NH_3 removal under higher loadings. It was also observed that the relative abundance of NO_2^- increased from 0.3 to 18% and that of NO_3^- decreased from 46 to 1%. Both phenomena indicated that the NOB activities were inhibited under high loadings (**Figure 4**). NH_4^+ and NO_2^-, which are the undissociated forms of NH_3 and HNO_2, are common inhibitors of nitrifying bacteria. The results in this study are consistent with other findings [28, 29]. NH_3 removal in a biotrickling filter is deteriorated due to the accumulation of NH_4^+ and NO_2^- which induces inhibitory effects on both AOB (*Nitrosomonas* sp.) and NOB (*Nitrobacter* sp.).

Among the four biofilters, the highest removal capacity (65%) was achieved in Reactor 2 (50% wood chips + 50% compost) under low loading rate (**Figure 2**). This is also demonstrated

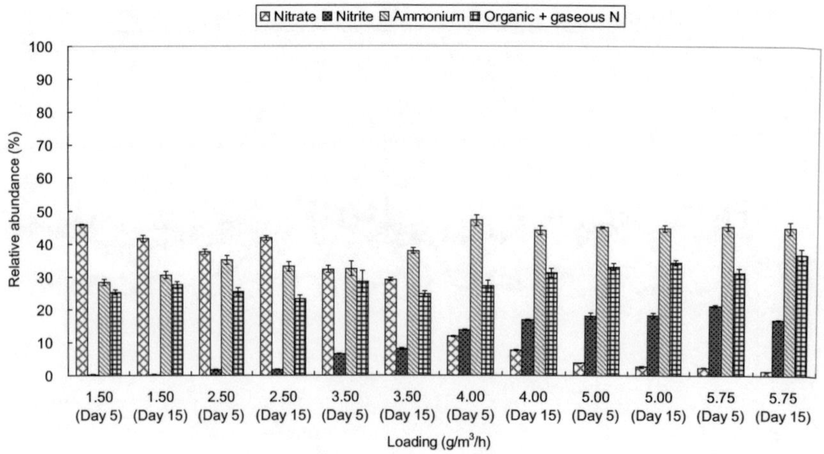

Figure 4. Mass balance analysis of metabolic products in Reactor 5.

by the pH profiles of each biofilter (**Figure 3**). At the low loading rate of 1.5 g m^{-3} h^{-1}, the treatment performance of other three biofilters (Reactors 1, 3, and 4) was around 17–20%. As the loading rates increased from 2.5 to 5.75 g m^{-3} h^{-1}, the trend of different reactor performances was similar to that at the loading rate of 1.5 g m^{-3} h^{-1}. The removal efficiency of Reactor 2 dropped significantly from 60 to 18.8%. For Reactors 1, 3 and 4, the removal efficiencies dropped substantially from around 21 to 5.5%.

With the exception of Reactor 2, adsorption and absorption were the main removal mechanisms for the other three biofilters. The dominant form of nitrogen in Reactors 1, 3, and 4 was NH_4^+. It was above 85% under all operating conditions, indicating that the absorption of NH_3 into the water layer was the main removal process. Therefore, poor treatment performance was achieved in Reactors 1, 3, and 4. In Reactor 2, the relative abundance of NH_4^+ was around 55% and remained stable under all operating conditions. However, the amount of NO_3^- and NO_2^- in Reactor 2 was 25% of the total nitrogen, which was about 3–10 times in Reactors 1, 3, and 4. Therefore, part of the NH_3 in Reactor 2 was removed by the activities of AOB and NOB through the oxidation of NH_3 into NO_2^- and NO_3^-. The quantity of NO_2^- increased from 5% to over 20% when the loading increased from 1.5 to 5.75 g m^{-3} h^{-1}, indicating that the relatively poor NH_3 removal capacity of Reactor 2 may be due to the inhibitory effects of high NH_3 concentration on the NOB community.

In Reactors 1, 3, and 4, the high pH values in the effluent implied the accumulation of ammonium and was mainly due to the overloading of ammonia. High pH and ammonium concentration inhibited the nitrification rates, especially for the activities of NOB. This consequently resulted in the significant increase of nitrite concentration [30].

In this study, Reactor 2 had a significantly higher ammonia removal efficiency compared with the other three biofilters, while Reactor 5 had the highest removal efficiency among the five reactors. It could be explained by the presence of a well-developed biofilm. Since Reactor 5 could provide a more favorable environment for bacterial growth, and the recirculation liquid could remove the toxic by-products, such as nitrite and nitrate, produced by nitrification. At the same time, the water layer outside the biofilm could provide a buffer capacity to absorb part of the ammonia for further bacterial activities. The length of start-up could also be shortened by the suitable inoculation of acclimated activated sludge bacteria in the reactor. Based on the findings in this experiment, the biotrickling operating mode was selected for detailed study.

3.4. Microbial community analysis

Figure 5a to **c** illustrates the electropherograms of the samples collected from Reactor 2, Reactor 5, and the recirculation of Reactor 5, after a 40-day operation, respectively. The relative diversity of the bacterial community is related to the number of peaks in the electropherogram. Meanwhile, the degree of abundance constituting each bacterial group is correlated to the intensity and area of the specific peak. Generally, the appearance or disappearance of peaks indicated changes in the bacterial community of the treatment system during the operation.

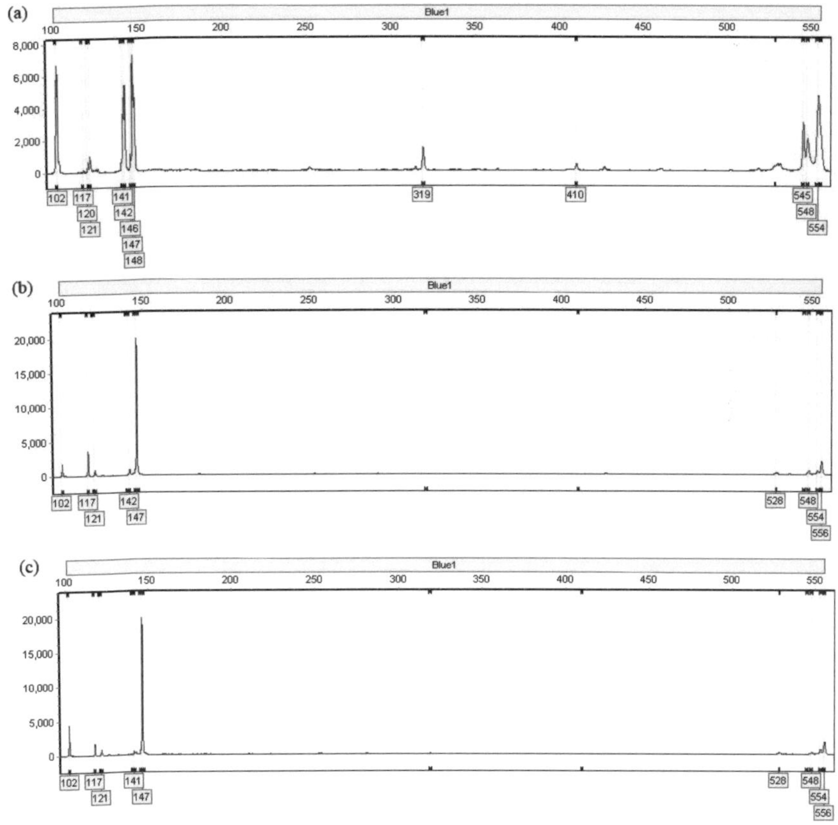

Figure 5. (a) Electropherogram of the sample collected from the compost-based biofilter (Reactor 2) during the preliminary study for the single NH_3 treatment. (b) Electropherogram of the sample collected from the biotrickling filter (Reactor 5) during the preliminary study for the single NH_3 treatment. (c) Electropherogram of the sample collected from the recirculation liquid in the biotrickling filter (Reactor 5) during the preliminary study for the single NH_3 treatment.

The results showed that a higher level of bacterial diversity was found in the compost biofilter (Reactor 2). However, the intensity of its peaks was less than that of the biotrickling reactor (Reactor 5). This indicates that the lower treatment performance of the compost biofilter may be due to the diminutive amount of bacteria developed in the biofilm. Perhaps, another possible reason would be that the compost biofilter was not inoculated with acclimated microorganisms from the activated sludge. Although the compost contains higher bacterial diversity, the indigenous bacteria may not grow well under the operating conditions. This results in a poor treatment performance and a longer start-up period.

Table 5 shows the relative abundance of the bacteria that was determined by the size of terminal restriction length (T-RF). The typical AOB, *Nitrosomonas europaea,* was found in both reactors. In contrast, the results showed a different composition of AOB in Reactors 2 and 5. It may be the reason for the varied treatment performance. A number of common bacteria in the activated sludge were also discovered in the community, including *Pseudomonas* sp. and *Bacillus* sp. Interestingly, a common sulfide-oxidizing bacteria (SOB), *Thiobacillus,* was found in the compost-based biofilter with a relatively high abundance (around 20%). It may be probably due to some H_2S or reduced sulfur in the compost which acted as the nutrients for the growth of SOB.

A similar bacterial population distribution was found in the medium and the recirculation liquid of the biotrickling system. This showed that the analysis of recirculation liquid is satisfactory for the bacterial community analysis of reactor operated in the performance study. This may be useful for the general monitoring because it is difficult to frequently collect the coal slag from the closed treatment system for the microbial study.

TF size (bp)	Possible bacteria	Compost	Coal slag	Recirculation liquid
Nitrifiers		Relative abundance (%)		
147	*Nitrosomonas, Nitrosospira*	21.3	59.3	58.7
554	*Nitrosococcus, Nitrosomonas*	20.6	3.6	5.0
Other microorganisms		Relative abundance (%)		
105	*Pseudomonas*	17.1	4.9	12.2
121	*Vibrio, Bacillus*	1.7	2.5	2.4
319	*Actinobacillus, Pseudomonas, Vibrio*	4.0	0.0	0.0
410	*Actinobacillus*	1.0	0.0	0.0
528	*Rhodopseudomonas*	0.0	1.9	1.6
548	*Lactobacillus, Pseudomonas, Thiocapsa*	16.4	3.7	2.4
554	*Thiobacillus*	20.6	3.6	5.0
556	*Bacillus*	0.0	12.2	12.1

Table 5. Terminal fragments and their corresponding bacteria.

http://dx.doi.org/10.5772/intechopen.68155

3.5. Performance of ammonia removal

The NH_3 removal efficiency in the biotrickling filter was investigated at the inlet concentrations in the range from 250 to 450 ppm and at a constant flow rate of 5 l min^{-1} for more than 9 months. The corresponding inlet loading rate was from 78.2 to 140.8 g m^{-3} h^{-1} under the operating conditions. Less than 2 days were required to adapt to the new operating condition for NH_3 treatment during each step increase of the mass loadings. Thereafter, the new steady state is also reached, and the pH of recirculation liquid was maintained between 6.5 and 7 in this experiment.

Figure 6 shows the NH_3 removal efficiency as a function of the inlet concentration. The results demonstrated that the bacterial community of AOB in the reactor provided a stable and consistent removal even at the beginning of the experiments. In general, the removal efficiency decreased with the increase of NH_3 concentration. A complete NH_3 removal was achieved during the operation at the inlet concentration of 250 ppm. No NH_3 (below the detection limit = 0.1 ppm) was consistently detected at the outlet. A superior treatment performance, higher than 98%, was achieved at the inlet concentration below 400 ppm. The removal efficiency remained above 99.9% at the concentration of 275 ppm (0.2 ppm detected at the outlet). This slightly decreased to 98.9% at the concentration of 350 ppm (4.6 ppm detected at the outlet). The outlet concentration significantly increased from 4.6 to 75.2 ppm when the NH_3 concentration increased from 350 to 450 ppm (removal efficiency dropped from 98.9 to 83.3%). The complete removal capacity and critical loading rate of the system were 90.5 g m^{-3} h^{-1} and 108.1 g m^{-3} h^{-1}, respectively, while the maximum elimination capacity was 118 g m^{-3} h^{-1}. **Table 6** shows the comparison of NH removal capacities in various biofiltration systems. The results of this study are relatively high in comparison with other studies [20, 26, 31].

When the loading rate was higher than the critical value, the biofilm would be completely saturated with NH_3 until a loading rate of 140.8 g m^{-3} h^{-1}. Moreover, the mass transfer limitation would inhibit the overall removal capacity at the loading rate less than 108.4 g m^{-3} h^{-1}

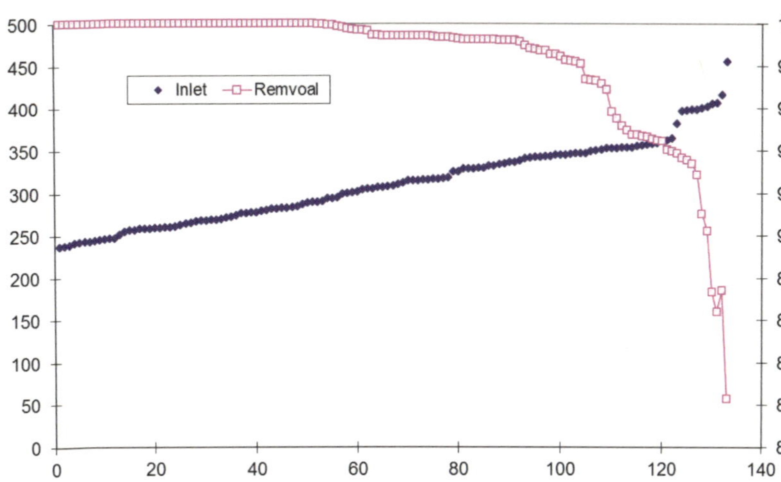

Figure 6. NH_3 removal efficiencies under various concentrations at the flow rate of 5 l min^{-1}.

Packing material	Microorganism	Critical loading (g m^{-3} h^{-1})	Max. elimination capacity (g m^{-3} h^{-1})	References
Biofilter				
Peat	Night soil sludge	30.0	41.7	Kim et al., [31]
Rock wool	Night soil sludge	33.8	50.0	Kim et al., [31]
Fuyolite	Night soil sludge	22.1	28.3	Kim et al., [31]
Ceramics	Night soil sludge	23.8	38.3	Kim et al., [31]
Fuyolite	*Vibrio alginolyticus*	93.0	114.0	Kim et al., [31]
Granulated sludge	Activated sludge	5.8	20.8	Gracian et al., [36]
Compost + bark + peat	Activated sludge	19.0	22.6	Choi et al., [37]
50% organic + 50% inorganic	Activated sludge	11.8	14.0	Choi et al., [36]
Compost + 20% perlite	Activated sludge	12.0	-	Chen et al., [26]
Sludge + 20% GAC	Activated sludge	10.1	-	Chen et al., [26]
Pall ring	Activated sludge	4.5	5.5	Kim et al., [20]
Biotrickling				
Ceramics	Soil	34.3	49.4	Kanagawa et al., [33]
Coal slag	Activated sludge	108.4	140.8	Present study

Table 6. Comparison of NH_3 removal capacities in various biofiltration systems.

[20, 32]. Authors reported that 4.5 s was required for the biodegradation of NH_3 by nitrifying bacteria [33]. In this study, the system was operated at EBRT of 8 s. This showed that the decrease of removal may be due to the mass transfer limitation from gas to liquid phases or the inhibitory effect of high NH_3 concentration on the oxidation activity of AOB.

At lower loading rates (i.e., 78 to 86 g m^{-3} h^{-1}), the main metabolite in the system was NO_3^-. This confirmed that the complete removal obtained was mainly contributed by the activities of autotrophic AOB and not by the physical absorption or adsorption. In fact, the organic and gaseous nitrogen was only around 5% under these conditions. The percentages of nitrite and ammonium increased directly proportional to the loading. At the highest loading rate, ammonium became the dominant by-product which accounted for more than 50%, while the removal efficiency was higher than 80%. Although a very high removal efficiency of NH_3 in the system was attained, a complete nitrification in the biofilm was never achieved due to the inhibitory effects of high NH_3 concentration.

4. Conclusions

The results of this study suggest that the biotrickling filter is a viable and effective method for the NH_3 removal. A comparatively short start-up of the system was accomplished within a 4-day operation. Excellent removal efficiency was achieved below the NH_3 loading rate at 108.4 g m^{-3} h^{-1} (i.e., 350 ppm) in this study, while the maximum elimination capacity was 118 g m^{-3} h^{-1}. Product analysis is allowed for the mass balances on nitrogen to identify the biodegradation processes that were active in the system. The main metabolites of NH_3 oxidation were ammonium, nitrate, and nitrite. No clogging and air channeling were observed during a long-term operation. For full-scale applications, odorous gas emitted from different industrial processes always contains high concentration of H_2S and low concentration of NH_3; the nitrogen metabolites available in the liquid may enhance the simultaneous treatment of NH_3 and H_2S from the waste gas stream in biotrickling filters. The acidic by-product (SO_4^{2-}) may be neutralized by the unionized ammonia, and a portion of sulfur compounds may act as nutrients for the growth of AOB and other bacteria; thus, the system could be easily maintained at a neutral range for a long-term operation.

Acknowledgements

This research was supported by the Research Grants Council of the Hong Kong SAR, China (No. 28300015 and 18202116); the Dean's Research Fund (FLASS/DRF/SFRS-8) and Internal Research Grant (RG 78/2015-2016) of the Education University of Hong Kong; and the Office of the Illinois Attorney General. The authors are also grateful for the financial support from the Joint Departmental Research Grant, University of Illinois at Urbana-Champaign.

Author details

Yiu Fai Tsang[1,2*], Ya-nan Wang[3], Huawei Wang[4,5], Yi Yang[1], Yuanhui Zhang[6] and Hong Chua[7]

*Address all correspondence to: tsangyf@eduhk.hk

1 Department of Science and Environmental Studies, The Education University of Hong Kong, Tai Po, New Territories, Hong Kong

2 Guizhou Academy of Sciences, Guiyang City, Guizhou Province, PR China

3 College of Environmental Science and Engineering, Tongji University, Shanghai, PR China

4 Xinjiang Key Laboratory of Environmental Pollution and Bioremediation, Xinjiang Institute of Ecology and Geography, Chinese Academy of Sciences, Urumqi, PR China

5 College of Environmental and Municipal Engineering, Qingdao University of Technology, Qingdao, PR China

6 Department of Agricultural and Biological Engineering, University of Illinois at Urbana-Champaign, USA

7 Faculty of Science and Technology, Technological and Higher Education Institute of Hong Kong, Tsing Yi, Hong Kong

References

[1] Bibbiani C, Russo C. Odour emission from intensive livestock production system: Approaches for emission abatement and evaluation of their effectiveness. Large Animal Review. 2012;18:135-138.

[2] Gabriel D, Maestre JP, Martin L, Gamisans X, Lafuente J. Characterisation and performance of coconut fibre as packing material in the removal of ammonia in gas- phase biofilters. Biosystems Engineering. 2007;97:481-490. DOI: 10.1016/j. biosystemseng.2007.03.038.

[3] Lin YH, Chen YP, Ho KL, Lee TY, Tseng CP. Large-scale modular biofiltration system for effective odor removal in a composting facility. Journal of Environmental Science and Health Part A-Toxic/Hazardous Substances and Environmental Engineering. 2013;48:1420-1430. DOI: 10.1080/10934529.2013.781898.

[4] Posmanik R, Gross A, Nejidat A. Effect of high ammonia loads emitted from poultry-manure digestion on nitrification activity and nitrifier-community structure in a compost biofilter. Ecological Engineering. 2014;62:140-147. DOI: 10.1016/j.ecoleng.2013.10.033.

[5] Taghipour H, Shahmansoury MR, Bina B, Movahdian H. Operational parameters in biofiltration of ammonia-contaminated air streams using compost-pieces of hard plastics filter media. Chemical Engineering Journal. 2008;137:198-204. DOI: 10.1016/j. cej.2007.04.015.

[6] Cohen Y Biofiltration – The treatment of fluids by microorganisms immobilized into the filter bedding materials: A review. Bioresource Technology. 2001;77:257-274. DOI: 10.1016/S0960-8524(00)00074-2.

[7] Kennes C, Rene ER, Veiga MC. Bioprocesses for air pollution control. Journal of Chemical Technology and Biotechnology. 2009;84:1419-1436. DOI: 10.1002/jctb.2216.

[8] Mudliar S, Giri B, Padoley K, Satpute D, Dixit R, Bhatt P, Pandey R, Juwarkar A, Vaidya A. Bioreactors for treatment of VOCs and odours–A review. Journal of Environmental Management. 2010;91:1039-1054. DOI: 10.1016/j.jenvman.2010.01.006.

[9] Pagans E, Font X, Sanchez A. Biofiltration for ammonia removal from composting exhaust gases. Chemical Engineering Journal. 2005;113:105-110. DOI: 10.1016/j.cej.2005.03.004.

[10] Tsang YF, Chua H, Sin SN, Chan SY. Treatment of odorous volatile fatty acids using a biotrickling filter. Bioresource Technology. 2008;99:589-595. DOI: 10.1016/j. biortech.200 6.12.032.

[11] Dorado A, Lafuente F, Gabriel D, Gamisans X. A comparative study based on physical characteristics of suitable packing materials in biofiltration. Environmental Technology. 2010;31:193-204. DOI: 10.1080/09593330903426687.

[12] Pagans E, Font X, Sanchez A. Adsorption, absorption, and biological degradation of ammonia in different biofilter organic media. Biotechnology and Bioengineering. 2007;-97:515-525. DOI: 10.1002/bit.21246.

[13] Anet B, Couriol C, Lendormi T, Amrane A, Le Cloirec P, Cogny G, Fillieres R. Characterization and selection of packing materials for biofiltration of rendering odourous emissions. Water Air and Soil Pollution. 224;7:1-13. DOI: 10.1007/s11270-013-1622-1.

[14] Yang L, Wang X, Funk TL, Gates RS. Biofilter media characterization and airflow resistance test. Transactions of the ASABE. 2011;54:1127-1136.

[15] Yang L, Wang X, Funk TL, Gates RS, Zhang Y. Transport and fate of nitrogen-containing compounds in gas-phase biofilters: A swing test to mitigate ammonia. Transactions of the ASABE. 2012:55:1951-1958.

[16] Chan SY, Tsang YF, Chua H, Sin SN, Cui LH. Performance study of vegetated sequencing batch coal slag bed treating domestic wastewater in suburban area. Bioresource Technology. 2008;99:3774-3781. DOI: 10.1016/j.biortech.2007.07.018.

[17] Chan SY, Tsang YF, Chua H. Domestic wastewater treatment using tidal-flow cinder bed with *Cyperus alternifolius*. Journal of Aquatic Ecosystem Health and Management Society. 2008;11:206-211. DOI: 10.1080/14634980802100022.

[18] Chan SY, Tsang YF, Cui LH, Chua H. Domestic wastewater treatment using batch-fed constructed wetland and predictive model development for ammonia removal. Process Biochemistry. 2008;43:297-305. DOI: 10.1016/j.procbio.2007.12.009.

[19] Chung YC, Ho KL, Tseng CP. Two-stage biofilter for effective NH_3 removal from waste gases containing high concentrations of H_2S. Journal of the Air & Waste Management Association. 2007;57:337-347.

[20] Kim JH, Rene ER, Park HS. Performance of an immobilized cell biofilter for ammonia removal from contaminated air stream. Chemosphere. 2007;68:274-280. DOI: 10.1016/j.chemosphere.2006.12.100.

[21] APHA. Standard methods for examination of water and wastewater. 2005. 21st ed. Washington, DC: APHA, AWWA, WPCF.

[22] Regan JM, Harrington GW, Noguera DR. Ammonia- and nitrite-oxidizing bacterial communities in a pilot-scale chloraminated drinking water distribution system. Applied and Environmental Microbiology. 2002;68:73-81. DOI: 10.1128/AEM.68.1.73-81.2002.

[23] Chung YC, Lin YY, Tseng CP. Removal of high concentration of NH_3 and coexistent H_2S by biological activated carbon (BAC) biotrickling filter. Bioresource Technology. 2005;96:1812-1820. DOI: 10.1016/j.biortech.2005.01.003.

[24] Duan H, Yan R, Koe LCC, Wang X. Combined effect of adsorption and biodegradation of biological activated carbon on H_2S biotrickling filtration. Chemosphere. 2007;66:1684-1691. DOI: 10.1016/j.chemosphere.2006.07.020.

[25] Duan H, Koe LCC, Yan R. Treatment of H_2S using a horizontal biotrickling filter based on biological activated carbon: reactor start-up and performance evaluation. Applied and Microbiology Biotechnology. 2005;67:143-149. DOI: 10.1007/s00253-004-1771-7.

[26] Chen YX, Yin J, Wang KW. Long-term operation of biofilters for biological removal of ammonia. Chemosphere. 2005;58:1023-1030. DOI: 10.1016/j.chemosphere.2004.09.052.

[27] Sheridan B, Curran T, Dodd V, Colligan J. Biofiltration of odour and ammonia from a pig unit-A pilot-scale study. Biosystems Engineering. 2002;82:441-453. DOI: 10.1006/bioe.2002.0083.

[28] Kim H, Xie Q, Kim YJ, Chung JS. Biofiltration of ammonia gas with sponge cubes coated with mixtures of activated carbon and zeolite. Environmental Technology. 2002;23:839-847. DOI: 10.1080/09593332308618355.

[29] Melse RW, Moi G. Odour and ammonia removal from pig house exhaust air using a biotrickling filter. Water Science and Technology. 2004;50:275-282.

[30] Anthonisen AC, Loehr RC, Prakasam TBS, Srinath EG. Inhibition of nitrification by ammonia and nitrous-acid. Journal Water Pollution Control Federation. 1976;48:835-852. DOI: 10.2307/25038971.

[31] Kim NJ, Hirai M, Shoda M. Comparison of organic and inorganic packing materials in the removal of ammonia gas in biofilters. Journal of Hazardous Materials. 2000;B72:77-90. DOI: 10.1016/S0304-3894(99)00160-0.

[32] Ottengraf SPP. Exhaust gas purification. In: Rehm HJ, Reed G, editors. Biotechnology. 1986. Vol. 8. Weinheim: VCH Verlagsgesellschaft. pp. 426-452.

[33] Kanagawa T, Qi HW, Okubo T, Tokura N. Biological treatment of ammonia gas at high loading. Waster Science and Technology. 2004;50:283-290.

[34] Amann RI, Binder RJ, Olson S, Chisholm SW, Devereux R, Stahl DA. (1990) Combination of 16S ribosomal RNA targeted oligonucleotide probes with flow cytometry for analyzing mixed microbial populations. Applied and Environmental Microbiology, 56, pp. 1919-1925.

[35] Choi JH, Kim YH, Joo DJ, Choi SJ, Ha TW, Lee DH, Park IH, Jeong YS. (2003) Removal of ammonia by biofilters: A study with flow-modified system and kinetics, Journal of the Air & Waste Management Association,53, pp. 92-101.

[36] Gracian C, Malhautier L, Fanlo JL, Cloirec PLe. (2002) Biofiltration of air loaded with ammonia by granulated sludge. Environmental Progress, 21, pp.237-245.

[37] Lin C, Stahl DA. (1995) Comparative analyses reveal a highly conserved endoglucanase in the cellulolytic genus Fibrobacter. Journal of Bacteriology, 177, pp. 2543-2549.

Chapter 5

Denitrification in the Presence of Chlorophenols: Progress and Prospects

Emir Martínez-Gutiérrez, Anne-Claire Texier,
Flor de María Cuervo-López and Jorge Gómez

Additional information is available at the end of the chapter

http://dx.doi.org/10.5772/intechopen.68860

Abstract

Diverse industrial effluents may contain recalcitrant compounds such as chlorophenols. Besides, excessive use of pesticides in agriculture is a major cause of the appearance of chlorophenols in surface and groundwater. To mitigate and control the effects of chlorophenols in the environment, various methods have been developed for their elimination. Biological processes represent a sustainable and economical alternative that can lead to the mineralization of chlorophenols and be effective for the removal of these pollutants from different water bodies, such as rivers, groundwater, and wastewater. Some studies have reported that chlorophenols mineralization and nitrate reduction may simultaneously be performed. Other works have suggested that a reductive dechlorination occurs such as the first step and later, the phenol formed is subsequently mineralized by denitrification. However, the published information can be confusing as the denitrifying process is often associated with the sole nitrate consumption without corroborating the total reduction of nitrate to N_2. Additionally, there are alternative systems that combine biological process with a chemical or electrochemical process for chlorophenols removal. This chapter focuses on the advances accomplished in the study of the removal of chlorophenols under denitrifying conditions with the aim of having a clearer panorama of the treatment alternatives that can be applied for treatment of this type of effluents.

Keywords: denitrification, chlorophenols, rates, anaerobic, combined systems

1. Introduction

Human activities generate effluents from production processes and domestic activities which may contain nitrogen and carbon pollutants. This pollution alters the global nitrogen and carbon cycles. Inorganic nitrogen is mainly present in the aqueous effluents such as nitrate, nitrite, and ammonium, causing serious problems to ecosystems and to public health.

These compounds can achieve high levels of toxicity to aquatic organisms and may promote the growth of aquatic plants, which accelerate the eutrophication process of water bodies [1]. The ingestion of nitrite and nitrate by infants can promote methemoglobinemia and the formation of nitrosamines, which might be carcinogens [2]. On the other hand, diverse industrial effluents may contain recalcitrant compounds such as chlorophenols, which are derivatives of phenol that contain one or more covalently bonded chlorine atoms. Chlorophenols have been utilized for wood preservation, as well as for manufacturing of pesticides, antiseptics, and dyes. However, the excessive use of pesticides in agriculture is one of the major causes of the appearance of chlorophenols in surface and groundwater [3]. Depending on their concentration, they can be toxic compounds, causing damage to the cell membranes as well as uncoupling oxidative phosphorylation [4].

To diminish the adverse effects of chlorophenols in the environment, various methods have been developed for their elimination, including physical, chemical, electrochemical, and biological processes. The first three methods appear to be faster, but everything indicates that they are expensive and generate collateral contamination, making them less environmentally friendly processes than the biological treatment. Biological processes represent a sustainable and cost-effective alternative that can lead to the mineralization of chlorophenols and can be effective for the removal of these pollutants from different water bodies, such as rivers, groundwater, and wastewater. Most of the information on disposal of chlorophenols under anaerobic conditions has been obtained under methanogenic conditions. There is evidence that shows that removal of chlorophenols by methanogenic microbial consortia is initiated by a reductive dechlorination and ends with the formation of methane and CO_2 [5], although more chlorinated chlorophenols are not always completely mineralized and less chlorinated compounds are obtained as end products [6]. Chlorophenol mineralization coupled to denitrification is still poorly documented. In this regard, there are few studies showing the possibility of chlorophenol consumption coupled to reduction of nitrate, although it is suggested that the pathway is different to reductive dechlorination [7]. Other studies suggested that reductive dechlorination occurs at first and later the phenol formed is subsequently mineralized by denitrification process [8]. However, the published information can be confusing, as the denitrifying process is often associated with the sole nitrate consumption without corroborating the total reduction of nitrate to N_2.

Considering that efficient removal of recalcitrant compounds such as chlorophenols requires detailed analysis, this chapter focuses on the advances accomplished in the study of the removal of chlorophenols under denitrifying conditions. First, the physicochemical characteristics of the chlorophenols that confer their recalcitrant and toxic properties are presented. Then, general aspects of denitrification, such as microbiology and biochemistry, as well as the influence of various environmental factors, are presented. In physiological terms, the elimination of chlorophenols under denitrifying conditions is discussed, presenting the different configurations of reactors studied, types of inoculum, as well as the different strategies used to increase their consumption. Finally, the recently studied systems that combine the biological process with a chemical or electrochemical process, in order to increase the consumption of chlorophenols without the generation of toxic waste, are also presented.

2. Physicochemical properties of chlorophenols

Chlorophenols are organochlorine compounds whose structure consists of a phenol and one or more chlorine atoms that are covalently bonded. In total, there are 19 types of chlorophenols differing from each other in the amount and position of the chlorine atoms. They can be subdivided into five groups: monochlorophenols, dichlorophenols, trichlorophenols, tetrachlorophenols, and pentachlorophenols. Most chlorophenols are solid at room temperature, with the exception of 2-chlorophenol (2-CP) which is liquid. They are compounds with strong odor and medicinal taste with very low organoleptic thresholds, being perceived in water at very small quantities (μg/L). Chlorophenols present high log Kow (octanol water partition coefficient) values and low solubility in water (**Table 1**). As chlorination level increases, their water solubility decreases and their acidity increases. Similarly, the log Kow also increases with the number of chlorine atoms, favoring their bioaccumulation [9]. Transport and transformation of chlorophenols in natural environments depend on pH, oxygen concentration, presence of other organic and inorganic substances, and temperature as well as their own structure [10].

Apparently, toxicity of chlorophenols is related to the degree of chlorination and the proximity of chlorine to the hydroxyl group. Chlorophenols with chlorine in the *ortho* position show lower toxicity than chlorophenols with the same number of chlorine in the *meta* or *para* position [11]. Toxicity of chlorophenols may also be related to their log Kow [12], as toxicity increases with a higher lipophilicity. Toxic effects of chlorophenols have been related to membrane destruction and inhibition of oxidative phosphorylation. This blockade of oxidative phosphorylation can occur by different ways: interfering with the release of hydrogen to the electron transport chain, inhibition of the transfer of electrons along the electron transport chain to oxygen, interference with the release of oxygen to the terminal electron carrier, or inhibition of the activity of adenosine triphosphate (ATP) synthase [11].

	2-CP	4-CP	2,4-DCP	2,4,6-TCP	2,3,5,6-TTCP	PCP
Molecular weight (g/mol)	128.56	128.56	163.0	197.45	197.45	266.34
Solubility at 20°C (g/L)	28	27	4.5	0.434	0.434	0.014
Density (g/cm^3)	1.262	1.2238	1.38	1.49	1.84	1.98
Log Kow	2.29	2.4	3.17	3.7	4.9	5.02
Vapor pressure at 20°C (mm Hg)	0.99	0.23	0.14	0.03	0.0059	0.0002
Melting point (°C)	9.4	42–44	42–43	69	115	191
Boiling point (°C)	174.9	217	210	246	288	309
pKa	8.56	9.18	7.68	6.0	5.5	5.01

CP: chlorophenol, DCP: dichlorophenol, TCP: trichlorophenol, TTCP: tetrachlorophenol, and PCP: pentachlorophenol.

Table 1. Physical and chemical properties of chlorophenols.

3. Microbiology of denitrification

In order to carry out denitrification, which is defined as the biological dissimilative transformation of nitrate (NO_3^-) or nitrite (NO_2^-) into molecular nitrogen (N_2) under anoxic conditions with energy conservation [13], an electron donor is required. Therefore, denitrifying microorganisms must have the ability for using nitrate or nitrite as electron acceptors to reduce them to molecular nitrogen. Organotrophic or autotrophic microorganisms are involved in denitrification process depending on their ability to use organic or inorganic compounds as electron sources, respectively. Their remarkable characteristic is their facultative anoxic respiration.

Distribution of denitrifying microorganisms in nature is ubiquitous. Organotrophic and autotrophic denitrifiers belong to α-, β-, γ- and ε-proteobacteria group which comprise both, Gram-negative and Gram-positive bacteria. Nevertheless, some members of Archae and Eukarya have shown the ability for reducing nitrate to N_2 [14, 15]. Most of organotrophic and autotrophic denitrifiers grow under neutral and mesophilic conditions [16, 17]. Organotrophic denitrifiers have been found in natural ecosystems as soil [18], surface water [19], groundwater, and sediments [20]; in wastewater treatment plants; and in different types of reactor configurations treating synthetic wastewater under organotrophic [21], autotrophic [22], or mixotrophic conditions, where mixtures of both organic and inorganic electron sources are present [22, 23]. For illustration purposes, several denitrifying microorganisms and their physiological characteristics are included in **Table 2**.

Group	Genus/species	Electron donor	Physiological characteristics	Reference
α-Proteobacteria	*Paracoccus/P. pantotrophus*	Organic and sulfur compounds, H_2	Organotrophic, sulfur and hydrogen autotrophic denitrification	[16, 24]
β-Proteobacteria	*Thiobacillus thiophillus*	Sulfide, sulfur	Sulfur autotrophic denitrification	[25]
	Azoarcus	Organic compounds	Organotrophic denitrification	[23, 26, 27]
	Thauera	Acetate, sulfide, H_2	Organotrophic, sulfur and hydrogen autotrophic denitrification	[21, 23]
	Acidovorax spp.	Glucose, acetate, H_2	Organotrophic and hydrogen autotrophic denitrification	[21, 28, 29]
	Flavobacterium spp.	Glucose, acetate	Organotrophic denitrification	[21, 28]
γ-Proteobacteria	*Pseudomonas sp.*	Organic compounds, H_2	Organotrophic and hydrogen autotrophic denitrification	[24]
	Acinetobacter sp.	H_2	Hydrogen autotrophic denitrification	[30]
	Aeromonas sp.	H_2	Hydrogen autotrophic denitrification	[31]
ε-Proteobacteria	*Sulfurimonas lithotrophicum*	Sulfur	Sulfur autotrophic denitrification	[32]
	Thiomicrospira CVO	Sulfur, H_2	Sulfur and hydrogen autotrophic denitrification	[33]

Table 2. Some denitrifying microorganisms and their physiological characteristics.

Genus *β-proteobacteria* has been found dominant in many denitrification systems [34]. *Thauera* is a dominant Gram-negative organotrophic bacterium belonging to *β-proteobacteria* which has been identified in wastewater treatment systems [35], in an integrated system of three-dimensional biofilm-electrode reactor and sulfur autotrophic denitrification (3DBER-SAD) under mixotrophic conditions [27]. *Thauera* has also been identified in sequential batch reactors where the heterotrophic and autotrophic denitrifying process was conducted [23] and in several denitrifying bioreactors under autotrophic conditions, suggesting its ability for autotrophic growth [23]. *Acidovorax* is a Gram-negative bacterium which has the ability of using both acetate and hydrogen for denitrification [29]. Denitrifying bacteria, similar to *Acidovorax* and *Azoarcus,* a facultatively anaerobic, mesophilic, and Gram-negative bacterium with the ability of growing with a variety of organic substrates [26], have been identified under mixotrophic denitrifying conditions [27]. Denitrifying species of *Acidovorax* spp. and *Flavobacterium* spp. have been detected in a soil column system amended with glucose [21]. Recently, the ability of *Pseudomonas sp.* C27 for conducting both organotrophic and autotrophic denitrification has been reported [22]. On the other hand, *Thiobacillus denitrificans,* an obligate autotrophy and facultative anaerobic bacterium, which can use elemental sulfur as an electron donor, has been isolated from natural environments, man-made environments [17], and denitrifying reactors operated under mixotrophic conditions [27].

4. Biochemical aspects

Irrespective of whether the organic or autotrophic process is conducted, the denitrification process has been described as a modular organization in which every biochemical reaction is catalyzed by a specific reductase [36]. These reactions occur when no oxygen is available and the environment becomes anoxic [37]. According to Mariotti [38], the denitrification process can be described as Eq. (1) indicates.

$$2NO_{3^-} + 10e^- + 12H^+ \rightarrow N_2 + 6H_2O \qquad \Delta G^{\circ\prime} = -1120.5\ \text{KJ/reaction} \tag{1}$$

This general equation can be decomposed into four enzymatic reactions. At first, nitrate is reduced to nitrite by nitrate reductase *(Nar)* (Eq. (2)). The reaction can take place in the cell membrane and periplasmic space. Affinity constant (*Km*) ranging from 0.15–15 mM and $\Delta G^{\circ\prime}$ of −163.2 KJ/reaction values have been reported for this reaction [39, 40]. UQH_2 corresponds to reduced ubiquinone, UQ to ubiquinone, c^{2+} to reduced cytochrome, and c^{3+} to oxidized cytochrome.

$$NO_{3^-} + UQH_2 \rightarrow NO_{2^-} + UQ + H_2O \tag{2}$$

A subsequent reduction of nitrite to nitric oxide is carried out by one of two nitrite reductases *(Nir, CuNir)* or the cytochrome cd_1 both located at the periplasmic space (Eq. (3) and (4)). *Km* values of 3.13–750 μM [41, 42] and 6–46 μM [39, 41] are reported for *Nir/CuNir* or *cd1,* respectively, whereas $\Delta G^{\circ\prime}$ of −73.2 KJ/reaction correspond to this stage.

$$(a)\ NO_2^- + Cu^{1+} + 2H^+ \rightarrow NO + H_2O + Cu^{2+} \quad (3)$$

or

$$(b)\ NO_2^- + c^{2+} + 2H^+ \rightarrow NO + H_2O + c^{3+} \quad (4)$$

Afterward, in the cell membrane, nitric oxide is reduced to nitrous oxide by the enzyme nitric oxide reductase (*Nor*) (Eq. (5)). Km values of 0.25–60 μM are reported for *Nor* enzyme with a ΔG°′ of −306.3 KJ/reaction [43, 44].

$$2NO + 2c^{2+} + 2H^+ \rightarrow N_2O + H_2O + 2c^{3+} \quad (5)$$

Finally, nitrous oxide is reduced to N_2 by the enzyme nitrous oxide reductase (*Nos),* which is located at the periplasmic space (Eq. (6)). *Km* values of 2–6 μM are reported for this enzyme with a ΔG°′ of −306.3 KJ/reaction [45].

$$N_2O + 2c^{2+} + 2H^+ \rightarrow N_2 + H_2O + 2c^{3+} \quad (6)$$

5. Denitrification and its environment

Denitrification performance is controlled by many environmental factors such as concentration, type and solubility of the substrate, C/N ratio, temperature, and pH, among other factors. These environmental variables determine the metabolic behavior, being the effect of each factor different on the biochemistry and physiology of the microorganisms [39, 46]. In this regard, experimental data have suggested that a C/N ratio close to the stoichiometric value is required for complete denitrification [47]. In this respect, many authors have made recommendations to adjust the C/N, S/N ratio for denitrification processes [36, 48]. Tiedje [49] observed that an excess of reducer source induced the reduction of nitrate to ammonium. Denitrification is an exergonic process where the energy formation depends on the type of reducer source. Degradation of monochlorophenols coupled to denitrification is also an exergonic process (**Table 3**). This makes denitrification a potential microbial biomass producer. Nonetheless, wastewater treatment should be a dissimilatory process where the pollutants might be essentially removed through catabolic processes.

Oxygen is generally considered as a denitrifying inhibitor [50]. Likewise, according to O_2 and nitrate potential redox, a competition effect can occur between these oxidants. It has been reported that nitrate could be reduced even in the presence of O_2 [51]. On the other hand, the denitrifying process can be carried out in a temperature range between 5 and 35°C. However, it has been observed that at low temperatures, the emissions of nitrous oxide increase whereas N_2 formation decreases [52].

pH is an independent variable that affects denitrification process at different levels [46, 53]. The common pH value employed for denitrification is around 7. At low pH values, an inhibition on the reduction of nitrous oxide occurs, causing an accumulation of nitrous oxide and a decrease

Compound	Equation	ΔG°′ (KJ/reaction)
Acetic acid	$CH_3COOH + 1.6NO_3- \rightarrow 2CO_2 + 0.8N_2 + 1.6OH^- + 1.2H_2O$	−843
Glucose	$C_6H_{12}O_6 + 4.8NO_3- \rightarrow 2.4N_2 + 6HCO_3- + 1.2H^+ + 2.4H_2O$	−2686
Phenol	$C_6H_6O + 5.6NO_3- + 0.2H_2O \rightarrow 2.8N_2 + 6HCO_3- + 0.4H^+$	−2818
Methanol	$CH_3OH + NO_3- \rightarrow 0.5N_2 + CO_2 + 2H_2O$	−582
p-Cresol	$C_7H_8O + 6.8NO_3- \rightarrow 3.4N_2 + 7HCO_3- + 0.2H^+ + 0.4H_2O$	−3422
Toluene	$C_7H_8 + 7.2NO_3- + 0.2H^+ \rightarrow 3.6N_2 + 7HCO_3- + 0.6H_2O$	−3524
Xylene	$C_8H_{10} + 8.4NO_3- + 0.4H^+ \rightarrow 4.2N_2 + 8HCO_3- + 1.2H_2O$	−4136
Sulfide	$S^{2-} + 2NO_3- + 4H^+ \rightarrow SO_4^{2-} + N_2 + 2H_2O$	−922
Monochlorophenol	$C_6H_5ClO + 5.2NO_3- + 1.4H_2O \rightarrow 2.6N_2 + 6HCO_3- + 1.8H^+ + Cl^-$	−2742

Table 3. Stoichiometric reactions of the denitrifying respiratory process using different electron sources and their ΔG°′ values (according to Cuervo-López et al. [36]).

in N_2 formation [54, 55]. Denitrification can also be influenced by the speciation and bioavailability of the chemical compounds used as reducer sources. Thus, physicochemical conditions must be controlled in order to have a faster and efficient denitrifying process.

6. Biodegradation of chlorophenols under denitrifying conditions

Chlorophenols are generally degraded under anaerobic conditions through the first reductive dechlorination step, which consists of the substitution of chlorine atoms by hydrogen atoms (Eq. (7)).

$$R-Cl + H_2 \rightarrow R-H + HCl \tag{7}$$

This stage is catalyzed by specific dehalogenases enzymes. The majority of the known reductive dehalogenases belong to the CprA/PceA family and contain one corrinoid and two iron-sulfur clusters as cofactors [56]. Reductive dechlorination requires the addition of electron donors. There are other cases in which chlorophenols are used as carbon and energy sources for microorganisms [57]. Under methanogenic conditions, mineralization of various chlorophenols to CO_2 and methane has been observed [5]. However, it is unclear if reductive dechlorination would be involved in the degradation of chlorophenols under denitrifying conditions. In fact, different pathways that do not involve the dechlorination reductive step have been suggested [7].

The study of chlorophenols under denitrifying conditions has been mainly evaluated using monochlorophenols. Chang et al. [58] used a biofilm to remove 2-CP under denitrifying conditions in batch cultures. They observed that the nitrate disappeared in 16 h, and there was a consumption of 2-CP. However, there was no formation of phenol in this period, suggesting that 2-CP was not dechlorinated in the presence of nitrate. Phenol was produced only after the disappearance of nitrate, suggesting that nitrate competed with 2-CP as an electron acceptor. A similar behavior was observed by Sanford and Tiedje [8], who evaluated, in serological bottles, the elimination of 2-CP in the presence of nitrate and acetate. They observed that the consumption of 2-CP was inhibited by the presence of nitrate and was only carried out when nitrate disappeared or was found in concentrations lower than 104 mg/L. Yu et al. [59] studied the effect of nitrate addition on the reductive dechlorination of pentachlorophenol (PCP) and found that low concentrations of nitrate (0–62 mg/L) can enhance reductive dechlorination of PCP, whereas high concentrations (310–1860 mg/L) provoke a contrary effect. Thus, reductive dechlorination could be carried out at low concentrations of nitrate. On the other hand, Häggblom et al. [60] studied the removal of three monochlorophenols in batch cultures under denitrifying conditions. Only 2-CP was eliminated in 110 days; nevertheless, they did not detect the formation of phenol as a product of reductive dechlorination. Bae et al. [7] also studied the elimination of monochlorophenols and dichlorophenols under denitrifying conditions in batch cultures, finding that 4-chlorophenol (4-CP), 2,4-dichlorophenol (2,4-DCP), and 2,6-dichlorophenol (2,6-DCP) were not biodegraded, whereas 2-CP and 3-chlorophenol (3-CP) were mineralized and the presence of nitrate was essential. The authors reported that 2-CP was oxidized to CO_2 under denitrifying conditions and suggested the presence of a population capable of eliminating 2-CP by a mechanism that does not involve reductive dechlorination.

As observed in **Table 4**, most of the studies with chlorophenols have been carried out in batch assays and only few types of reactors have been evaluated under denitrifying conditions. Moussavi et al. [61] evaluated the elimination of 2-CP in a granular anoxic baffled reactor (AnBR) increasing the concentration of 2-CP up to 500 mg/L without affecting the efficiency of 2-CP removal, so this could be a feasible process at low cost. Wang et al. [62] evaluated the removal of PCP in a packed reactor with corncob as both carbon source and biofilm support and obtained efficiencies of nitrate and PCP removal above 90%.

In conclusion, mineralization of chlorophenols coupled to denitrification is rarely documented as the total oxidation of chlorophenols to CO_2 and reduction of nitrate to N_2 have not been corroborated. The available information is controversial as several works evidenced that the presence of nitrate inhibits the transformation of chlorophenols [8, 63], while other authors indicate that reductive dechlorination can be carried out at low concentrations of nitrate [59]. In fact, other studies evidenced that mineralization of chlorophenols is linked to denitrification, and the presence of nitrate was necessary for the biodegradation [7, 64]. In addition, the denitrifying process is often evaluated by the sole nitrate consumption without verifying its total reduction to N_2. Therefore, it is necessary to carry out more studies which evaluate the process through response variables such as removal efficiencies, yields of product formation, and rates in order to characterize and better understand the process.

6.1. Strategies for improving the consumption of chlorophenols

It has been pointed out that the main difficulty for the elimination of chlorophenols is the strong stability that the carbon-halogen bond of the aromatic compound confers to its structure [67]. Thus, in many cases, the biodegradation is slow. Several strategies for increasing the consumption

Chlorophenol (mg/L)	Type of reactor	Inoculum	Electron donor	Removal efficiency of CPs	Removal efficiency of nitrate	Products	Reference
2-CP (12.8)	Batch (serum bottle)	Sediment	$Na_2S.9H_2O$	—	—	—	[60]
2-CP, 3-CP, or 4-CP (25.7)	Batch	Soil	Acetate, volatile fatty acids	—	—	Phenol, benzoate	[8]
2-CP (12.8)	Batch	Acclimated sludge	$Na_2S.9H_2O$	—		——	[7]
3-CP (2.0–5.2)	Up-flow columns	Activated sludge	$Na_2S.9H_2O$	27–100%	—	Phenol, benzoate	[65]
2-CP(25)	Batch (gas-permeable silicone membrane bioreactor)	Hydrogenotrophic biofilm (acclimated)	H_2	Around 100%	Around 100%	Phenol	[58]
4-CP, 2,4 DCP (5)	Sequencing batch reactors	Acclimated biomass	Milk powder plus yeast extract	—	—	—	[66]
PCP (5 mg/L)	Laboratory-scale reactor packed	Biofilm	Corncob	40–91%	98%	3-CP, phenol	[62]
2-CP (50–500)	Anoxic baffled reactor	Activated sludge (enrichment)		>99%	—	—	[61]
PCP (1–5)	Batch (serum bottles)	Soil	Lactic acid	Around 100%	Around 100%	—	[59]

Table 4. Biodegradation of different chlorophenols under denitrifying conditions.

of chlorophenols have been proposed, although most of them have been conducted under aerobic and anaerobic conditions and in minor proportions underdenitrifying conditions.

Some strategies have been proposed to increase the efficiency and/or rate of chlorophenols consumption. These include the sludge adaptation to pollutants, the use of genetically modified microorganisms, and the addition of alternative carbon sources [68]. It has been also suggested that the addition of readily oxidized carbon sources could exert various beneficial effects, such as decreasing toxicity, acting as an enzyme-inducing agent, or providing reducing power for the consumption of recalcitrant organic compounds [69–71]. Furthermore, Puyol et al. [72] observed accumulation of different intermediates depending on the co-substrate used. When methanol, ethanol, or volatile fatty acids were used as co-substrates, 4-chlorophenol was accumulated while 2,4-dichlorophenol was accumulated when lactate was used as the co-substrate.

Under denitrifying conditions, Hu et al. [66] found that the presence of co-substrates caused a significant decrease in the degradation rate of 4-chlorophenol (by 4 times) while the biodegradation rate of 2,4-dichlorophenol increased by 4.2 times. Therefore, it could be said that the use of co-substrates does not always have a positive effect on the biodegradation of recalcitrant compounds. The compounds used as co-substrates include compounds of easy oxidation and compounds with a structure similar to chlorophenols. Regarding this, Martínez-Gutiérrez et al. [73] evaluated the effects of phenol and acetate on the mineralization of 2-CP by a denitrifying sludge in batch assays. When phenol was used as a co-substrate, the specific rate of 2-CP consumption increased by 2.6 times, regarding to a control assay without co-substrate. When acetate was used, the specific rate of 2-CP consumption increased by 9 times, suggesting that the addition of co-substrates is a good alternative for improving the biodegradation of chlorophenols. These results also suggest that the effects of co-substrates addition depend on several factors: type of both the co-substrate and chlorophenol employed, inoculum source, and experimental conditions.

7. Coupled systems for chlorophenol degradation

Recently, other strategies have been developed for the elimination of recalcitrant compounds using systems that combine advanced oxidation (AOP) or electrochemical processes with biological processes. Daghio et al. [74] evaluated the degradation of toluene using bio-electrochemical reactors obtaining a current power of 431 mA/m^2. Yeruva et al. [75] evaluated the integration of a sequencing batch reactor (SBR) and a bio-electrochemical treatment system (BET) for the treatment of petrochemical wastewater under anoxic conditions, obtaining high degradation and power generation (17.12 mW/m^2). The application of an electrochemical treatment can diminish the time required for the treatment of chlorinated pesticides in the biological process [76]. A sequential biological advanced oxidation process was used for the degradation of 2,4-dichlorophenol, consisting of an up-flow anaerobic sludge blanket (UASB) reactor and a $UV/H_2O_2/TiO_2$ system, obtaining 52.7% of degradation in only 6 h [77]. However, the degradation of chlorophenols with nitrate using combined systems has been scarcely evaluated. In this sense, Arellano-González et al. [78] evaluated an electrochemical-biological combined system where the reductive dechlorination was carried out in an ECCOCEL-type (Pd-Ni/Ti electrode) reactor that achieved 100% transformation of 2-CP into phenol. Then, the phenol formed was mineralized by a biological denitrification process. The total time required for 2-CP conversion into CO_2 was 7.5 h.

8. Perspectives

Biodegradation processes of chlorophenols have been studied extensively because they are more economical and friendly environmental processes in comparison with physicochemical, AOP, and electrochemical processes. The information presented in this review shows that denitrification might be an efficient biological process for the treatment of effluents contaminated with nitrogen and chlorophenols. It has been also reported that biological processes may achieve the complete removal of many types of chlorophenols under aerobic and anaerobic conditions, but they do not always lead to mineralization. It is crucial considering that biodegradation processes can generate more toxic and recalcitrant intermediates than the original pollutant, and the partial oxidation of recalcitrant molecules should be prevented, favoring their mineralization. In this review, it is shown that recent experimental evidences demonstrated the possibility to use denitrification for 2-CP mineralization associated with the reduction of nitrate to nitrogen gas. These results suggest that denitrification might be used for the mineralization of chlorophenols producing CO_2 and N_2 as final products and obtaining high removal efficiencies. However, more studies on chlorophenols biodegradation by denitrifying processes are needed, especially with mixtures of chlorophenols. More studies on physiological, kinetic, and biochemical aspects of denitrification are also required to identify the limiting steps of the biodegradation metabolic pathways and to better understand how controlling denitrifying processes in bioreactors without the formation of undesirable by-products.

Another important aspect is that it has been shown that chlorophenols biodegradation by denitrifying microorganisms is very slow. As a consequence, the application of denitrification processes for chlorophenols removal is still limited, requiring very long acclimation and retention times, especially for the treatment of wastewaters contaminated with high chlorophenol concentrations. Different treatment alternatives have been proposed in order to increase the rate and efficiency of chlorophenol consumption and among them are adaptation to the pollutants, utilization of genetically modified microorganisms, and addition of alternative sources of energy. However, in spite of the addition of co-substrates, the time required for complete mineralization of chlorophenols can be still very long compared to those obtained in physicochemical processes. In recent years, there have been proposals for coupling oxidation processes (AOP or electrochemical) to biological processes such as denitrification to combine benefits of both types of treatment and establish more efficient, more rapid, less expensive, and environmentally friendly treatment trains for degradation of recalcitrant compounds in wastewater. One alternative is the pretreatment of chlorophenols containing effluents through chemical or electrochemical processes to make them more easily degradable in a sequencing denitrifying biological treatment. Recent results showed that times can be considerably reduced for the complete mineralization of 2-CP in an electrochemical-biological combined system, where an electrocatalytic dehydrogenation process (reductive dechlorination) was coupled to a biological denitrification process in sequential ECCOCEL-type (Pd-Ni/Ti electrode) and rotating cylinder denitrifying reactors. The total time required for 2-CP mineralization in the combined electrochemical-biological process was close to the previously reported times for electrochemical and AOP processes, but in this case, an efficient process was obtained without accumulation of by-products or generation of excessive energy costs due to the selective electrochemical pretreatment. This study showed that the use of electrochemical reductive pretreatment combined with denitrification could be a promising technology for the removal of recalcitrant molecules, such as chlorophenols, from wastewater by more efficient, rapid, and environmentally friendly processes. However, more studies are required in order to get an insight about the denitrification of electrochemically pretreated effluents in different combined systems, different configurations of reactors, and in the presence of different mixtures of chlorophenols and types of co-substrates.

Author details

Emir Martínez-Gutiérrez[1]*, Anne-Claire Texier[2], Flor de María Cuervo-López[2] and Jorge Gómez[2]

*Address all correspondence to: emir.martinez@inecol.mx

1 CONACYT-Institute of Ecology, Veracruz, México

2 Biotechnology Department, Autonomous Metropolitan University-Iztapalapa, Mexico City, México

References

[1] Laws EA. Aquatic Pollution–An Introductory Test. USA: Interscience; 1993. p. 55

[2] Forman D. Nitrate exposure and human cancer. In: Bogardi I, Kuzelka R, editors. Nitrate Contamination, RD NATO ASI Series. G30 Berlin, Germany: Springer-Verlag; 1991. pp. 281–288. DOI: 10.1007/978-3-642-76040-2_20

[3] Olaniran AO, Igbinosa EO. Chlorophenols and other related derivatives of environmental concern: Properties, distribution and microbial degradation processes. Chemosphere. 2011;**83**:1297–1306. DOI: 10.1016/j.chemosphere.2011.04.009

[4] Bradberry SM, Proudfoot AL, Vale JA. Poisoning due to chlorophenoxy herbicides. Toxicological Reviews. 2004;**23**:65–73. DOI: 10.2165/00139709-200423020-00001

[5] Boyd S, Shelton D. Anaerobic biodegradation of chlorophenols in fresh and acclimated sludge. Applied and Environmental Microbiology. 1984;**47**:272–277. DOI: 0099-2240/84/020272-06$02.00/0

[6] Garibay-Orijel C, Ríos-Leal E, García-MenaJ, Poggui-Varaldo M. 2,4,6-Trichlorophenol and phenol removal in methanogenic and partially-aerated methanogenic conditions in fluidized bed reactors. Journal of Chemical Technology and Biotechnology. 2005;**80**(10):1180–1187. DOI: 10.1002/jctb.1313

[7] Bae H, Yamagishi T, Suwa Y. Evidence for degradation of 2-chlorophenol by enrichment cultures under denitrifying conditions. Microbiology. 2002;**148**:221–227. DOI: 10.1099/00221287–148-1-221

[8] Sanford R, Tiedje J. Chlorophenol dechlorination and subsequent degradation in denitrifying microcosms fed low concentration of nitrate. Biodegradation. 1996;**7**:425–434. DOI: 10.1007/BF00056426

[9] Chiou CT, Schmedding DW, Manes M. Improved prediction of octanol-water partition coefficients from liquid-solute water solubilities and molar volumes. Environmental Science and Technology. 2005;**39**(22):8840–8846. DOI:10.1021/es050729d

[10] Czaplicka M. Sources and transformations of chlorophenols in natural environment. Science of the Total Environment. 2004;**322**:21–39. DOI: 10.1016/j.scitotenv.2003.09.015

[11] Pepelko WE, Gaylor DW, Mukerjee D. Comparative toxic potency ranking of chlorophenols. Toxicology and Industrial Health. 2005;**21**:93–111. DOI: 10.1191/0748233705th204oa

[12] Smith S, Furay VJ, Layiwola PJ, Menezes–Filho JA. Evaluation of the toxicity and quantitative structure/activity relationships (QSAR) of chlorophenols to the copepodid stage of a marine copepod (*Tisbe battagliai*) and two species of benthnic flatfish, the flounder (*Platichthys flesus*) and sole (*Solea solea*). Chemosphere. 1994;**28**(4):825–836. DOI: 10.1016/0045-6535(94)90234-8

[13] Zumft WG. Cell biology and molecular basis of denitrification. Microbiology and Molecular Biology Reviews. 1997;**61**:533–616

[14] Shoun H, Kim DH, Uchiyama H, Sugiyama J. Denitrification by fungi. FEMS Microbiology Letters. 1992;**94**:277–281. DOI: 10.1111/j.1574-6968.1992.tb05331.x

[15] Hayatsu M, Tago K, Saito M. Various players in the nitrogen cycle: Diversity and functions of microorganisms involved in nitrification and denitrification. Soil Science and Plant Nutrition. 2008;**54**:33–45. DOI: 10.1111/j.1747-0765.2007.00195.x

[16] Robertson LA, Kuenen. *Thiosphera panthotropha* gen. nov. sp. nov., a facultatively anaerobic, facultatively autotrophic sulfur bacterium. Microbiology. 1983;**129**:2847–2855. DOI:10.1099/00221287-129-9-2847

[17] Shao M, Zhang T, Fang HH. Sulfur-driven autotrophic denitrification: diversity, biochemistry, and engineering applications. Applied Environmental and Biotechnology. 2010;**88**:1027–1042. DOI: 10.1007/s00253-010-2847-1

[18] Morris JT, Whiting GJ, Chapelle FH. Potential denitrification rates in deep sediments from the southeastern coastal plain. Environmental Science and Technology. 1988;**22** (7):832–836

[19] Bernot MJ, Dodds WK, Gardner WS, McCarthy MJ, Sobolev D, Tank JL. Comparing denitrification estimates for a Texas estuary by using acetylene inhibition and membrane inlet mass spectrometry. Applied and Environmental Microbiology. 2003;**69**:5950–5956. DOI: 10.1128/AEM.69.10.5950-5956.2003

[20] Torrentó C, Urmeneta J, Otero N, Soler A, Viñas M, Cama J. Enhanced denitrification in groundwater and sediments from a nitrate-contaminated aquifer after addition of pyrite. Chemical Geology. 2011;**287**(1-2):90–101. DOI: 10.1016/j.chemgeo.2011.06.002

[21] Calderer M, Marti V, de Pablo J, Guivernau M, Prenafeta-Boldú FX, Viñan M. Effects of enhanced denitrification on hydrodinamics and microbial community structure in a soil column system. Chemosphere. 2014;**111**:112–119. DOI: 10.1016/j.chemosphere.2014.03.033

[22] Gou H, Chen C, Lee DJ, Wang A, Gao D, Ren N. Coupled carbon, sulfur and nitrogen cycles of mixotrophic growth of *Pseudomonas* sp. C27 under denitrifying sulfide removal conditions. Bioresource Technology. 2014;**171**:120–126. DOI: 10.1016/j.biortech.2014.08.035

[23] Xu G, Peng J, Feng C, Fang F, Chen S, Xu Y, Wang X. Evaluation of simultaneous autotrophic and heterotrophic processes and bacterial community structure analysis. Applied Microbiology and Biotechnology. 2015;**99**:6527–6536. DOI: 10.1007/s00253-015-6532-2

[24] Szekeres S, Kiss I, Kalman M, Soares MI. Microbial population in a hydrogen-dependent denitrification reactor. Water Research. 2002;**36**(16):4088–4094. DOI: 10.1016/S0043-1354(02)00130-6

[25] Kellerman C,Griebler C. *Thiobacillus thiophilus*, sp. nov., a chemolithoautotrophic, thiosulfate-oxidizing bacterium isolated from contaminated aquifer sediments. International Journal of Systematic Evolution Microbiology. 2009;**59**(3):583–588. DOI: 10.1099/ijs.0.002808-0

[26] Junghare M, Patil Y, Schink B. Draft genome sequence of a nitrate-reducing, o-phthalate degrading bacterium, *Azoarcus* sp strain PA01T. Standards in Genomic Sciences. 2015;**10**:90. DOI:10.1186/s40793-015-0079-9

[27] Hao R, Meng C, Li JJ. Impact of operation condition on the denitrifying bacterial community structure in a 3DBER-SAD reactor. Journal of Industrial Microbiology and Biotechnology. 2017;**44**:9–21. DOI:10.1007/s10295-016-1853-4

[28] Osaka T, Yoshie S, Tsuneda S, Hirata A, Iwami N, Inamori W. Identification of acetate-or-methanol-assimilating bacteria under nitrate-reducing conditions by stable-isotope probing. Microbial Ecology. 2006;**52**(2):253–266. DOI: 10.1007/s00248-006-9071-7

[29] Vasiliadou IA, Pavlou S, Vayenas DV. Dynamics of chemostat with three competitive hydrogen oxidizing denitrifying microbial populations and their efficiency for denitrification. Ecological Modelling. 2009;**220**(8):1169–1180. DOI: 10.1016/j.ecolmodel.2009.02.009

[30] Mao Y, Xia Y, Zhang T. Characterization of *Thauera*-dominated hydrogen-oxidizing autotrophic denitrifying microbial communities by using high-throughput sequencing. Bioresource Technology. 2013;**128**:703–710. DOI: 10.1016/j.biortech.2012.10.106

[31] Wang H, He Q, Chen D, Wei L, Zou Z, Zhou J, Yang K, Zhang H. Microbial community in a hydrogenotrophic denitrification reactor based on pyrosequencing. Applied Microbiology and Biotechnology. 2015;**99**(24):10829–10837. DOI: 10.1007/s00253-015-6929-y

[32] Nakagawa S, Takai K, Inagaki F, Hirayama H, Nunoura T, Horikoshi K, Sako Y. Distribution phylogenetic diversity and physiological characteristics of epsilon-proteobacteria in a deep-sea hydrothermal field. Environmental Microbiology. 2005;**7**(10):1619–1632. DOI: 10.1111/j.1462-2920.2005.00856.x

[33] Gevertz D, Telanj AV, Voordouw G, Jenneman GE. Isolation and characterization of strains CVO and FWKOB, two novel nitrate reducing, sulfide-oxidizing bacteria isolated from oil field brine. Applied Environmental Microbiology. 2000;**66**(6):2491–2501. DOI: 10.1128/AEM.66.6.2491-2501.2000

[34] Yao S, Chen Q, Borthwick A. Enrichment and characterization of a bacteria consortium capable of heterotrophic nitrification and aerobic denitrification at low temperature. Bioresource Technology. 2013;**127**:151–157. DOI: 10.1016/j.biortech.2012.09.098

[35] Liu BB, Mao YJ, Bergaust L, Bakken LR, Frostegard A. Strains in the genus *Thauera* exhibit remarkably different denitrification regulatory phenotypes. Environmental Microbiology. 2013;**15**(10):2816–2828.DOI: 10.1111/1462-2920.12142

[36] Cuervo-López FM, Martínez-Hernández S, Texier A-C, Gómez J. Denitrification for waste water treatment. In: Cervantes F, editor. Environmental Technologies to Treat Nitrogen Pollution. Principles and Engineering. London, UK, International Water Association Publishing; 2009. pp. 41–65

[37] Berks BC, Ferguson SJ, Moir JWB, Richardson DJ. Enzymes and associated electron transport system that catalyse the respiratory reduction of nitrogen oxides and oxyanions. Biochemical et Biophysica Acta. 1995;**1232**(3):97–133. DOI: 10.1016/0005-2728 (95)00092-5

[38] Mariotti A. Denitrification in groundwaters, principles and methods for its identification: A review. Journal of Hydrology. 1986;**88**:1–23

[39] Betlach MR, Tiedje JM. Kinetic explanation for accumulation of nitrite, nitric oxide and nitrous oxide during bacterial denitrification. Applied and Environmental Microbiology. 1981;**42**:1074–1084

[40] Song B, Palleroni NJ, Kerkhof LJ, Haggblom MM. Characterization of halobenzoate-degrading, denitrifying *Azoarcus* and *Thauera* isolates and description of *Thauera chlorobenzoica* sp.nov. International Journal of Systematic and Evolutionary Microbiology. 2001;**51**:589–602. DOI: 10.1099/00207713-51-2-589

[41] Gudat GC, Singh J, Wharton DC. Cytochrome oxidase from *Pseudomonas aeruginosa*. Purification and some properties. Biochimica et Biophysica Acta. 1973;**292**(2):376–390. DOI: 10.1016/0005-2728(73)90044-3

[42] Michalski W, Nicholas DJD. Molecular characterization of copper containing nitrite reductase from *Rhodopseudomonas sphaeroides* form sp. denitrificans. Biochimica et Biophysica Acta. 1985;**828**:130–137.DOI: 10.1016/0167-4838(86)90146-9

[43] Carr GC, Ferguson SJ. The nitric oxide reductase of *Paracoccus denitrificans*. Biochemical Journal. 1990;**269**:423–429. DOI: 10.1042/bj2690423

[44] Fujiwara T, Fukumori Y. Cytochrome cb-type nitric oxide reductase with cytochrome c oxidase activity from *Paracoccus denitrificans* ATCC35512. Journal Bacteriology. 1996;**178**:1866–1871. DOI: 10.1128/jb.178.7.1866-1871.1996

[45] Berks BC, Baratta D, Richardson BJ, Ferguson SJ. Purification and characterization of a nitrous oxide reductase from *Thiosphaera pantotropha*. European Journal of Biochemical. 1993;**212**:467–476. DOI: 10.1111/j.1432-1033.1993.tb17683.x

[46] Saleh-Lakha S, Shannon KE, Henderson SL, Goyer C, Trevors KT, Zebarth BJ, Burton DL. Effect of pH and temperature on denitrification gene expression and activity in *Pseudomonasmandelii*. Applied and Environmental Microbiology. 2009;**75**:3903–3911. DOI: 10.1128/AEM.00080-09

[47] Elefsiniotis P, Li D. The effect of temperature and carbon source on denitrification using volatile fatty acids. Biochemical Engineering Journal. 2006;**28**:148–155. DOI: 10.1016/j.bej.2005.10.004

[48] Cervantes F, Monroy O, Gomez J. Influence of ammonium on the performance of a denitrifying culture under heterotrophic conditions. Applied Biochemistry and Biotechnology. 1999;**81**:13–21. DOI: 10.1385/ABAB:81:1:13

[49] Tiedje JM. Ecology of denitrification and dissimilation nitrate to ammonium. In: Zendher AJB, editor. Environmental Microbiology of Anaerobes. New York: Wiley; 1988. pp. 179–244

[50] Hernandez D, Rowe JJ. Oxygen regulation of nitrate uptake in denitrifying *Pseudomonas aeruginosa*. Applied and Environmental Microbiology. 1987;**53**:745–750. DOI: 0099-2240/87/040745-06$02.00/0

[51] Ferguson SJ. Denitrification and its control. Antonie van Leeuvenhoek. 1994;**66**:89–110. DOI: 10.1007/BF00871634

[52] Dobbie KE, Smith KA. The effects of temperature, water-filled pore space and land use on N_2O emissions from an imperfectly drained gleysol. European Journal of Soil Science. 2001;**52**(4):667–673. DOI: 10.1046/j.1365-2389.2001.00395.x

[53] Knowles R. Denitrification. Microbial Reviews. 2001;**46**:43–70. DOI: 0146-0749/82/010043-28$02.00/0

[54] Thomsen JK, Geest T, Cox RP. Mass spectrometric studies of the effect of pH on the accumulation of intermediates in denitrification by *Paracoccus denitrificans*. Applied and Environmental Microbiology. 1994;**60**:536–541

[55] Li YM, Li J, Zheng GH, Luan J-F, Fu QS, Gu G-W. Effects of the COD/NO_3^--N ratio and pH on the accumulation of denitrification intermediates with available pyridine as a sole electron donor and carbon source. Environmental Technology. 2008;**29**:1297–1306. DOI: 10.1080/09593330802379672

[56] Bisaillon A, Beaudet R, Lépine F, Déziel E, Villemur R. Identification and characterization of novel CprA reductive dehalogenase specific to highly chlorinated phenols from *Desulfitobacterium hafniense* strainPCP-1. Applied Environmental Microbiology. 2010;**76**(22):7536–7540. DOI: 10.1128/AEM.01362-10

[57] Field JA, Sierra-Alvarez R. Microbial degradation of chlorinated phenols. Reviews in Environmental Science and Biotechnology. 2008 **7**(3):211–241. DOI: 10.1007/s11157-007-9124-5

[58] Chang C-C, Tseng SK, Chang C-C, Ho C-H. Reductive dechlorination of 2-chlorophenol in a hydrogenotrophic, gas permeable, silicone membrane bioreactor. Bioresource Technology. 2003;**90**:323–328. DOI: 10.1016/S0960-8524(03)00149-4

[59] Yu H-Y, Wang Y-K, Chen P-C, Li F-B, Chen M-J, Hu M, Ouyang X. Effect of nitrate addition on reductive transformation of pentachlorophenol in paddy soil relation to iron

(III) reduction. Journal of Environmental Management. 2014;**132**:42–48. DOI: 10.1016/j.jenvman.2013.10.020

[60] Häggblom MM, Rivera MD, Young LY. Influence of alternative electron acceptors on the anaerobic biodegradability of chlorinated phenols and benzoic acids. Applied and Environmental Microbiology. 1993;**59**:1162–1167. DOI: 0099-2240/93/041162-06$02.00

[61] Moussavi G, Ghodrati S, Mohseni-Bandpaei A. The biodegradation and COD removal of 2-chlorophenol in a granular anoxic baffled reactor. Journal of Biotechnology. 2014;**184**:111–117. DOI: 10.1016/j.jbiotec.2014.05.010

[62] Wang X, Xing L, Qui T, Han M. Simultaneous removal of nitrate and pentachlorophenol from simulated groundwater using a biodenitrification reactor packed with corncob. Environmental Science and Pollution. 2013;**20**(4):2236–2243. DOI: 10.1007/s11356-012-1092-9

[63] D'Angelo EM, Reddy KR. Aerobic and anaerobic transformation of pentachlorophenol in wetlands soils. Soil Science Society of America Journal. 2000;**64**(3):933–943. DOI:10.2136/sssaj2000.643933x

[64] Jianlong W, Horan N, Stentiford ED, Yi Q. Anaerobic degradation of 4-chlorophenol under denitrifying conditions with nitrate as the only electron acceptor. Toxicological and Environmental Chemistry. 2001;**80**:209–216. DOI:10.1080/02772240109359011

[65] Bae H, Yamagishi T, Suwa Y. Developing and sustaining 3-chlorophenol-degrading populations in up-flow column reactors circum-denitrifying conditions. Applied Microbiology and Biotechnology. 2002b;**59**:118–124. DOI: 10.1007/s00253-002-0974-z

[66] Hu Z, Ferraina RA, Ericson JF, Smets BF. Effect of long-term exposure, biogenic substrate presence, and electron acceptor conditions on the biodegradation of multiple substituted benzoates and phenolates. Water Research. 2005;**39**:3501–3510. DOI: 10.1016/j.watres.2005.06.009

[67] Farrel A, Quilty B. The enhancement of 2-chlorophenol degradation by a mixed microbial community when augmented with *Pseudomonas putida* CP1. Water Research. 2002;**36**:2443–2450. DOI: 10.1016/S0043-1354(01)00481-X

[68] Fakhruddin ANM, Quilty B. The influence of glucose and fructose on the degradation of 2-chlorophenol by *Pseudomonas putida*. World Journal of Microbiology and Biotechnology. 2005;**21**(8):1541–1548. DOI: 10.1007/s11274-005-7580-z

[69] Perkins PS, Komisar SJ, Puhakka JA, Ferguson FJ. Effects of electron donors and inhibitors on reductive dechlorination of 2,4,6-trichlorophenol. Water Research. 1994;**28**:2101–2107. DOI: 10.1016/0043-1354(94)90020-5

[70] Chaudhuri BK, Wiesmann U. Enhanced anaerobic degradation of benzene by enrichment of mixed microbial culture and optimization of culture medium. Applied Microbiology and Biotechnology. 1995;**43**:178–187. DOI: 10.1007/BF00170641

[71] Martínez-Hernández S, Olguín EJ, Gómez J, Cuervo-López FM. Acetate enhances the specific consumption rate of toluene under denitrifying conditions. Archives of Environmental Contamination and Toxicology. 2009;**57**:679–687. DOI: 10.1007/s00244-009-9321-z

[72] Puyol D, Mohedano AF, Sanz JL, Rodriguez JJ. Comparison of UASB and EGSB performance on the anaerobic biodegradation of 2,4-dichlorophenol. Chemosphere. 2009;**76** (9):1192–1198. DOI: 10.1016/j.chemosphere.2009.06.015

[73] Martínez-Gutiérrez E, González-Márquez H, Martínez-Hernández S, Texier A-C, Cuervo-López FM, Gómez J. Effect of phenol and acetate addition on 2-chlorophenol consumption by a denitrifying sludge. Environmental Technology. 2012;**33**:1375–1382. DOI: 10.1080/09593330.2011.627882

[74] Daghio M, Vaipoulou E, Patil SA, Suárez-Suárez A, Head IM, Franzetti A, Rabaey K. Anodes stimulate anaerobic toluene degradation via sulfur cycling in marine sediments. Applied and Environmental Microbiology. 2016;**82**:297–307. DOI: 10.1128/AEM.02250-15

[75] Yeruva DK, Jukuri S, Velvizhi G, Kumar AN, Swamy YV. Integrating sequencing batch reactor with bio-electrochemical treatment for augmenting remediation efficiency of complex petrochemical wastewater. Bioresource Technology. 2015;**188**:32–43. DOI: 10.1016/j.biortech.2015.02.014

[76] Fontmorin J-M, Fourcade F, Geneste F, Floner D, Huguet S, Amrane A. Combined process for 2,4-cychlorophenoxyacetic acid treatment coupling of an electrochemical system with biological treatment. Biochemical Engineering Journal. 2013;**70**:17–22.DOI: 10.1016/j.bej.2012.09.015

[77] Dixit A, Tirpude AJ, Mungray AK, Chakraborty M. Degradation of 2,4 DCP by sequential biological-advanced oxidation process using UASB an $UV/TiO_2/H_2O_2$. Desalination. 2011;**272**:265–269. DOI: 10.1016/j.desal.2011.01.035

[78] Arellano-González MA, González I, Texier A-C. Mineralization of 2-chlorophenol by sequential electrochemical reductive dechlorination and biological processes. Journal of Hazardous Materials. 2016;**314**:181–187. DOI: 10.1016/j.jhazmat.2016.04.048

Chapter 6

Nitrate Removal from Groundwater with Membrane Bioreactor

Marjana Simonič, Andreja Goršek and
Aleksandra Petrovič

Additional information is available at the end of the chapter

http://dx.doi.org/10.5772/intechopen.68718

Abstract

The aim of this study is to model the denitrification process performed in a membrane bioreactor (MBR). The research was carried out using a modified Zenon ZeeWeed 10 MBR system. The membrane module consisted of submerged hollow-fibre membrane with a pore size of 0.04 μm and an active area of 0.93 m^2. The concentration of nitrate in drinking water was (70 ± 2) mg/L NO_3^-. During the experiment, we maintained a constant concentration level of activated sludge at approximately 0.76 g/L under anoxic conditions. Sugar was added to the activated sludge as a source of carbon. The Monod kinetic parameters were estimated based on the experimental data numerical interpolation. Afterwards, a dynamic simulation with known parameters was carried out, and the time dependence of the substrate and biomass concentration was studied. We developed a model based on actual substrate outlet concentration. In addition, the time required to reach a steady state was estimated.

Keywords: denitrification, groundwater, membrane bioreactor, dynamic concentration profile

1. Introduction

Nitrate and nitrite removal from water is necessary because of the harmful effects of nitrates on human health, such as methaemoglobinemia (blue-baby syndrome) [1–3], nitrosamines and nitrosamides [4].

During the biological process of denitrification, nitrate is microbiologically reduced over nitrite to molecular nitrogen (N_2) [1, 5]. The efficiency of biological removal of nitrate depends on different types of carbon sources [6, 7], various types of microorganisms [6, 8] and different operational parameters such as carbon to nitrogen (C/N) ratios [2, 9], temperature [3, 10–12], pH [3, 10], dissolved oxygen [13, 14] and mixed liquor suspended solids (MLSSs) [15, 16]. Furthermore, it also depends on the amount of substrate and heterotrophic yield [17]. Denitrification may be inhibited by higher levels of nitrate and nitrite [10, 18, 19], which can directly affect microbial growth. Reaction rates and efficiencies are sensitive to dissolved oxygen [11]; anoxic growth reaction especially can be inhibited [17]. The advantages of heterotrophic denitrification are, on the one hand, the high denitrifying rates; on the other hand, one of the greater weaknesses is that the residual carbon sources can cause many problems during drinking water treatment [3]. For growth under anoxic conditions, heterotrophic denitrifiers require a specific source of organic carbon, such as methanol [1, 2, 6, 20], ethanol [2, 15, 21], acetate [7, 10], glucose [7, 9, 20], glycerol [20] and acetic acid [20], whilst the application of sucrose is relatively rare and has only been mentioned in a few articles [2]. Gómez et al. [2] studied the effectiveness of three selected carbon sources (sucrose, ethanol and methanol) on submerged filters for the removal of nitrate from contaminated groundwater (100 mg/L NO_3^-). Greater biomass production was observed with sucrose, compared with ethanol and methanol. Fernández-Nava et al. [4] examined the properties of saccharose-rich residue (from the production of soft drinks) in the process of denitrification. Crude syrup as a C source was used in another study performed by Lee and Welander [6]. Sison et al. [22] used sucrose in the process of denitrification by biological granular-activated carbon. The influent NO_3-N concentration was 80 mg/L (C/N ratio 1.88:1), and the average denitrification efficiency achieved 84–89%. During the study, when the C/N ratio increased from 1.5 to 2.5, removal efficiency increased up to 95% [23]. Besides the influence of C sources, the investigations focused on different types of denitrification (hydrogenotrophic [19, 24], autotrophic [25], heterotrophic [25]), membrane bioreactor (MBR) configurations [16], carbon to nitrogen (C/N) ratio [15, 25–27], and the removal of pesticides [27]. Moreover, studies were carried out on hydraulic retention time [28–30], concentration of mixed liquor suspended solids (MLSSs) [15, 16, 30], mathematical modelling of MBR [31], optimisation of the energy demand [29], trihalomethane formation potential [11, 21, 28, 30] and the inhibition of nitrite [10, 19]. The first commercial-scale biological drinking water denitrification plant utilising hydrogen was introduced at Rasseln in Germany [19, 24]. However, the MBR system is, in general, less commonly used for drinking water treatment. Nitrate removal from contaminated groundwater, drinking water and surface water has been studied by using extractive MBRs [31, 32], ion-exchange MBRs [16], gas-transfer MBRs [16], pressure-driven MBRs [15, 16, 28] and other known hybrid systems [11, 25]. The Zenon ZW 10 membrane bioreactor was first used in the denitrification of drinking water sources in 2005 [26].

Miscellaneous models for describing the process kinetics have been studied so far (e.g. the Haldane model and Michaelis-Menten kinetics). The performance of a special bacterial culture (*Aphelenchus avenae*) was investigated using different carbon sources, such as ethanol, methanol, sodium acetate, glucose and poly(ε-caprolactone), within the batch biological denitrification system [33]. The most commonly used relationship describing microbial growth is Monod kinetics [33–35]. This mechanism is also used to describe heterotrophic denitrification [19]. There are several factors affecting microbial growth and its kinetics: pH [10, 36], temperature [12, 36, 37], dissolved oxygen [11, 36], type of substrate [2, 9], microbial population [12, 37, 38], type of water source [37] and the presence of nitrite [10]. The temperature dependence of the growth rate can generally be described using the Arrhenius relationship [39]. Microbial growth, Monod kinetics and the influences of different physico-chemical factors on the denitrification process have been extensively investigated in recent papers [34, 35, 39]. In an experiment by Ravindran et al. [11], mixed batch bioreactor studies were performed to evaluate the denitrification kinetics of groundwater. Ethanol was added as the external carbon source. A sensitivity analysis was performed in order to determine which biokinetic parameter had the greatest influence on effluent substrate concentration. The results obtained showed that biokinetic coefficients vary significantly with any changes in the MLSS concentrations of groundwater.

Studying biokinetic coefficients is important to obtain more information about the cell growth and utilisation of substrate, which then helps to better understand the denitrification process. A literature review shows that there is a lack of information related to the determination of kinetic coefficients for drinking water denitrification treatment by MBR using sugar as a C source.

The purpose of our research was to develop a kinetic model in order to describe microbial growth during the drinking water denitrification process using MBR. A kinetic study was conducted by assuming Monod kinetics to be appropriate for describing substrate consumption at constant biomass concentration. Firstly, the basic kinetic parameters, such as specific growth rate of biomass, substrate half-saturation constant and the yield coefficient, were determined based on experimental data. Furthermore, dynamic simulation was performed based on calculated kinetic parameters. With the dynamic concentration profiles, the time dependence of the substrate and biomass concentrations can be followed, and the time required to reach steady state can be estimated.

2. Materials and methods

During this study, denitrification was carried out in a modified Zenon ZW 10 MBR, which can be described as a continuous stirred-tank reactor (CSTR) with recycle capability. The microbial growth rate was expressed using Monod kinetics [34, 35, 39].

In our case, biomass was absent from the influent and effluent of the bioreactor, so the system behaved as a closed system (although the circulation of biomass within the reactor still existed). The fact that the increase of biomass was very low had to be taken into account during the calculations, in which the increase of biomass was neglected. Because of the biomass characteristics, it was assumed that the mode of MBR operation would be close to the model of mixed flow bioreactor under steady state. The substrate dynamic profiles could be described with the equations for the continuous stirred-tank reactor (CSTR) with recycle [40]. The mass balances for substrate and biomass [36, 40], provided a basis for determining a kinetic model regarding drinking water denitrification (Eqs. (1)–(26)). The balance of biomass was obtained in two ways: firstly, by using the equations for a continuous stirred-tank reactor with recycling, and secondly, by the equations for the reactor without recycling. During the testing of the second method, we assumed that the biomass concentration in the circulation is equal to the concentration of biomass in the reactor and would thus produce the same result in both cases, namely that the specific growth rate of microorganisms is equal to the dilution rate.

2.1. Mass balance of biomass

Depending on the biomass and the mode of operation, the equations for the CSTR in steady state [36, 40] can be used (Eqs. (1)–(9)):

$$\text{Input} - \text{Output} + \text{Generation} = \text{Accumulation} \tag{1}$$

For each part of above Eq. (1), the following can be written:

$$q_D\gamma_D - q_X\gamma_X + \mu\gamma_X V = (\mathrm{d}\gamma_X V/\mathrm{d}t) \tag{2}$$

Eq. (2) can be written in the form:

$$q_D\gamma_D - q_X\gamma_X + \mu\gamma_X V = (\mathrm{d}\gamma_X/\mathrm{d}t)V + (\mathrm{d}V/\mathrm{d}t)\gamma_X \tag{3}$$

For the continuous stirred-tank reactor, $(\mathrm{d}V/\mathrm{d}t) = q_D - q_X = 0$, and from this it follows that $q_D = q_X = q$; therefore, it can be written and referred to as

$$q(\gamma_D - \gamma_X) + \mu\gamma_X V = (\mathrm{d}\gamma_X/\mathrm{d}t)V/ : V \tag{4}$$

And then

$$q/V(\gamma_D - \gamma_X) + \mu\gamma_X = \mathrm{d}\gamma_X/\mathrm{d}t \tag{5}$$

The *quotient* of the inlet flow rate and bioreactor volume can be expressed as, $D = q/V$ (h^{-1}). Dynamic changes in the biomass concentration over the time can be written as follows:

$$D(\gamma_D - \gamma_X) + \mu\gamma_X = \mathrm{d}\gamma_X/\mathrm{d}t \tag{6}$$

By considering that the inlet mass concentration of biomass is zero ($\gamma_D = 0$) and at steady state $\mathrm{d}\gamma_X/\mathrm{d}t = 0$, Eq. (6) can be expressed as

$$-D\gamma_X + \mu\gamma_X = 0 \tag{7}$$

And

$$D = \mu \tag{8}$$

Specific growth rate of biomass, μ, can be expressed as [35, 36]

$$\mu = \mu_{Max}\gamma_S/(K_S + \gamma_S)) \tag{9}$$

At high substrate concentrations ($\gamma_S >> K_S$), a zero-order kinetic model is usually used and at low-substrate concentrations, first-order dependence can be applied [10, 39].

The specific growth rate of active biomass is a result of the endogenous decay of active biomass (microbial death) reduced for the coefficient k_d (h^{-1}) [36]. Thus, Eq. (6) can be rearranged into Eq. (10):

$$D(\gamma_D - \gamma_X) + (\mu - k_d)\gamma_X = d\gamma_X/dt \tag{10}$$

And further:

$$d\gamma_X/dt = D(\gamma_D - \gamma_X) + (\mu_{Max}\gamma_S/(K_S + \gamma_S) - k_d)\gamma_X \tag{11}$$

Similarly, the mass balance of biomass for CSTR with recycling [36, 40], can be written according to Eqs. (12)–(20):

$$\text{Input} - \text{Output} + \text{Generation} = \text{Accumulation} \tag{12}$$

In steady state: Accumulation = 0

Similarly, Eq. (3), the next Eq. (13) can be written as

$$q_D\gamma_D + q_R\gamma_R - (q_D + q_R)\gamma_X + \mu\gamma_X V = 0 \tag{13}$$

Since $q_D\gamma_D = 0$ (biomass concentration at the inflow is zero), therefore

$$q_R\gamma_R - (q_D + q_R)\gamma_X + \mu\gamma_X V = 0 \tag{14}$$

If Eq. (14) is divided by V and afterwards by γ_X, then we obtain the following expression:

$$q_R\gamma_R/(V\gamma_X) - q_D/V - q_R/V + \mu = 0 \tag{15}$$

The dilution rate is the reciprocal value of residence time, $D = q_o/V$, therefore

$$\mu = D + q_R/V - q_R\gamma_R/(V\gamma_X) \tag{16}$$

With the introduction of parameter, $a = q_R/q_o$, we obtain Eq. (17):

$$\mu = D + aq_o/V - (aq_o\gamma_R/(V\gamma_X) \tag{17}$$

And if the quotient γ_R/γ_X is replaced by parameter b, then Eq. (17) can be rewritten as

$$\mu = D + aD - aDb \tag{18}$$

Since in our bioreactor, there was no barrier (or cell separator) which could lead to changes in the concentrations, we assumed that the mass concentrations of biomass in the recycle and in the reactor are equal $\gamma_R = \gamma_X$. This leads to the assumption that $b = 1$ and from this it follows that the specific growth rate of the biomass is equal to the dilution rate, Eqs. (19) and (20).

$$\mu = D + aD - aD \tag{19}$$

And

$$\mu = D \tag{20}$$

In this way, the same final expression for biomass, as with the equations for the continuous stirred-tank reactor, was obtained.

2.2. Mass balance of substrate

Regarding the substrate, the equations for CSTR with recycle [36, 40], were adequate (Eqs. (21)–(26)):

$$\text{Input} - \text{Output} - \text{Consumption} = \text{Accumulation} \tag{21}$$

In steady state: Accumulation = 0

For the substrate according to Eq. (21), the following expression can be written:

$$q_D \gamma_{S,D} - q_D \gamma_S - (1/y_{X,S})\mu\gamma_X V = 0 \tag{22}$$

If Eq. (22) is divided by V and by considering that $q_o/V = D$, we obtain:

$$D\gamma_{S,D} - D\gamma_S - (1/y_{X,S})\mu\gamma_X = 0 \tag{23}$$

The yield coefficient can be determined according to Eq. (24), which describes the mass balance of substrate in the steady state.

$$D(\gamma_{S,D} - \gamma_S) = (1/\gamma_{X,S})\mu\gamma_X \tag{24}$$

Since during the consumption of the substrate and thus in the production of biomass only active biomass is involved, the variable w_x is introduced into Eq. (24) representing the percentage of active biomass:

$$D(\gamma_{S,D} - \gamma_S) = (1/\gamma_{X,S})\mu w_X \gamma_X \tag{25}$$

In the literature [8, 38, 41], information may be found about the proportion of active biomass, depending on a number of factors. The viability of biological sludge can be expressed as the active bacterial concentration per unit mass of volatile suspended solids [41].

Dynamic changes in the substrate concentration over time are displayed by the following Eq. (26):

$$D(\gamma_{S,D} - \gamma_S) - (1/\gamma_{X,S})\mu_{max}\gamma_X\gamma_S/(K_S + \gamma_S) = d\gamma_S/dt \tag{26}$$

2.3. Membrane bioreactor

Experiments were performed using the modified Zenon ZeeWeed 10 membrane bioreactor (MBR). The denitrification process was carried out under anoxic conditions in a reactor volume of 60 L. The average operating temperature within the reactor was 26.3°C and the pH value within the range of (8.7–9.4). Variations in operating temperatures were a result of changes in the external temperatures. The membrane module consisted of a submerged hollow-fibre membrane with a pore size of 0.04 μm and a 0.93 m^2 active area. The process scheme for the drinking water treatment within the modified MBR is shown in **Figure 1** and the ultrafiltration (UF) membrane specifications are presented in **Table 1**.

The groundwater used for the study was spiked with sodium nitrate in concentration (70 ± 2) mg/L NO_3^-. The membrane bioreactor (**Figure 2**) was inoculated with biomass sludge from an existing wastewater treatment plant. During the experiment, we maintained a constant

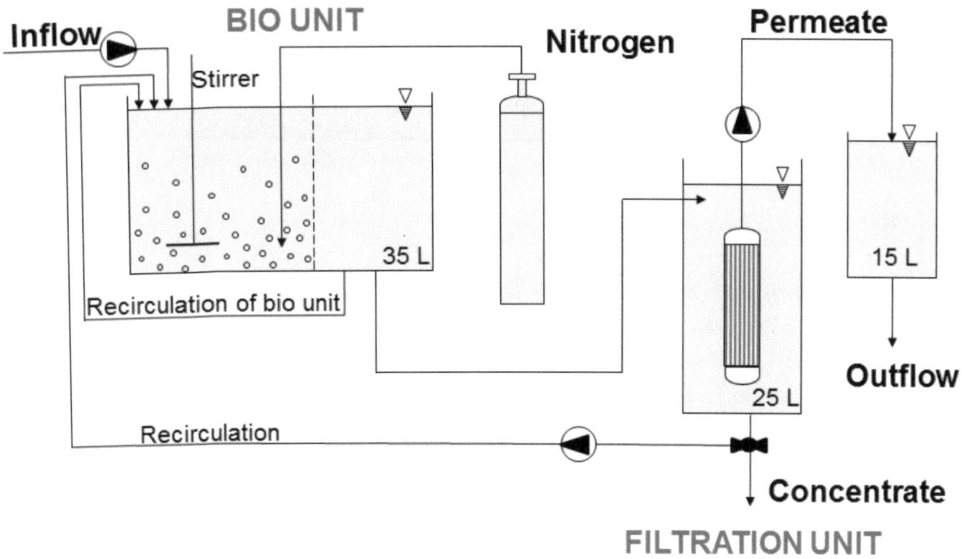

Figure 1. The process scheme for groundwater treatment with the modified MBR.

Specifications	Description
Type of membrane	Hollow fibre (HF)
Material	Polyvinylidene difluoride (PVDF)
Surface properties	Neutral, hydrophilic
Nominal membrane area	0.93 m^2
Pore size	0.04 μm
Max. temperature	40°C
pH range	5–9
Max. trans-membrane pressure	62 kPa
Max. pressure of backpulse	55 kPa
Max. capacity of process pumps	1.4 L/min

Table 1. UF membrane specifications.

Figure 2. Modified Zenon ZW 10 membrane module during the treatment.

concentration level of activated sludge at approximately 0.76 g/L. Anoxic conditions were provided by using nitrogen. Sugar was added to the activated sludge as a source of carbon. Inlet mass concentration of substrate was 0.1126 g/L. Based on previous papers [2, 22, 23] and our previous investigations, the appropriate value for the C/N ratio was 3:1.

A series of experiments were performed in order to follow the influence of drinking water flow rates (dilution rates) on the outlet's substrate concentration. The flow rate of the feed was increased stepwise, from 10 up to 170 mL/min. At each flow rate (or dilution rate), sufficient time was ensured to establish a steady state.

2.4. Analytical methods

Before, during, and after treatment of the drinking water, the following physico-chemical parameters were monitored: chemical oxygen demand (COD), content of nitrate ions NO_3^- and the mixed liquor suspended solids (MLSS). In addition, flow and circulation of water were monitored. Sugar concentration in the effluent was determined indirectly by measuring the chemical oxygen demand.

3. Results and discussion

Firstly, Monod kinetics parameters were determined.

During the experiment, we tried to maintain a constant concentration level of activated sludge within the reactor, 0.76 g/L expressed as MLSS and C/N ratio of 3:1. During the experiment, sugar concentration in the inflow was constant throughout all series. Such conditions allowed an average nitrate removal efficiency of 87%.

γ_S (mg/L)	D (h^{-1})
0	0
0.51	0.01
0.77	0.02
0.79	0.03
0.96	0.04
1.07	0.05
1.12	0.08
3.10	0.12
7.02	0.17

Table 2. Experimentally determined substrate mass concentration versus the dilution rate.

Determination of the kinetic parameters was based on the experimental values of outlet substrate mass concentration γ_S and the calculated dilution rates (D). The γ_S is expressed as chemical oxygen demand (COD). The data are gathered in **Table 2**.

Because of the low increment of biomass, changes in its concentration were negligible; therefore, MBR operation mode was close to that of a mixed flow bioreactor under steady state. For this reason, Eq. (8) could be adopted.The curve $D = f(\gamma_S)$ was plotted when compiling this equation, as shown in **Figure 3**. Subsequently, numerical interpolation using MATLAB software was performed.

The following results for Monod kinetics parameters were obtained: maximum specific growth rate of biomass, $\mu_{max} = 0.31\ h^{-1}$ ($7.4\ d^{-1}$) and the half-saturation constant (as COD) $K_S = 5.4$ mg/L, both with $R^2 = 0.94$.

In the existing literature, there is a lack of information regarding the Monod parameters for drinking water denitrification, and it was impossible to find data relating the value of maximum specific growth rate and the half-saturation constant for systems similar to ours.

The yield coefficient was determined in the next step of our study and afterwards dynamic simulation was performed. The value of the yield coefficient was computed according to Eq. (25). This equation considered whether in the consumption of the substrate and thus during the production of biomass, only the active part of the biomass is involved. Sears et al. [8] reported that under typical operating conditions the microbial fraction of the activated sludge flocs represents approximately 40% by weight, whilst Chung and Neethling [41] reported that only 5–10% of the total volatile suspended solids represented active bacterial biomass. Similar values for MBR processes have been reported, namely that an active fraction of biomass [38] is between 4 and 7%. Based on these data, an active fraction of biomass (w_x) in our research was set at 5%. Numerical interpolation of experimental results (by the method of least squares) was performed in order to determine the yield coefficient ($Y_{X/S}$). The calculated value of the yield coefficient was ($Y_{X/S}$) = 0.35 ($R^2 = 0.94$), which meant that approximately 35% of biomass was produced regarding the consumed substrate.

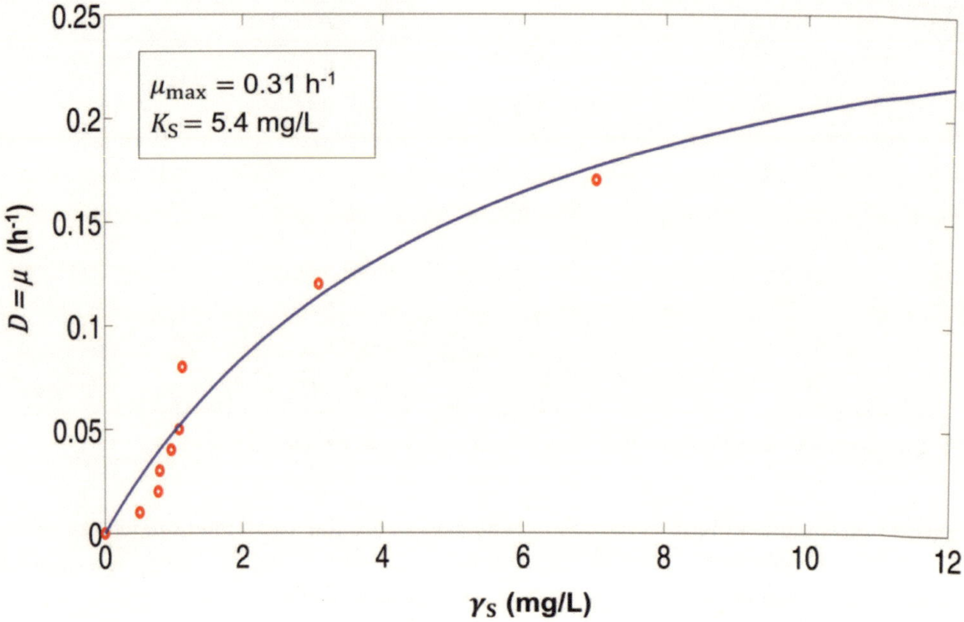

Figure 3. Specific growth rate of biomass as a function of substrate concentration at the outflow.

To date, no data for yield coefficient have been available for drinking water denitrification by MBR using sugar as the C source. The heterotrophic yield coefficient of activated sludge bacteria provides information about the biodegradability studies of chemical compounds and is important for modelling processes [42]. The removal of nitrate depends on the amount of substrate used and the heterotrophic yield [17]. In addition, product formation and yield coefficient are affected by temperature [36]. Lee and Welander [6], in their study on the effects of different carbon sources on respiratory denitrification, concluded that the carbon source had a significant influence on the denitrification rate, denitrification yield and the composition of the microflora. The growth yield for saccharose-rich crude syrup obtained during this study was within the range of 0.26–0.35 g TSS/g COD removed. The yield coefficient of aerobic organism growth using glucose was typically from 0.4 to 0.6, whilst the anaerobic growth was less efficient and the yield coefficient was reduced substantially [36].

3.1. Dynamic simulation

Dynamic simulation was performed based on the results obtained for μ_{max}, K_S and $Y_{X/S}$. Using dynamic simulation by means of a software program, the time required to establish a steady state was estimated and the impact of the dilution rate on the concentration profiles of substrate and biomass was studied. The equations applied to this were: Eq. (11), which provides the dynamic changes of the biomass concentration over time, and Eq. (26), which describes the dynamic changes of the substrate concentration over time. The dilution rate varied from 0.1 up to 5 d^{-1}. The value for the specific endogenous decay rate for the heterotrophic biomass was determined at k_d = 0.05 d^{-1}. Dynamic simulation was performed according to the proposed model by anticipating two different outlet substrate concentrations at the start of an operation (at the time of zero): first, the value of γ_S close to zero (software allowed a minimal value 0.001 g/L) and second, the actual γ_S is 0.1126 g/L. Dynamic concentration profiles are shown in **Figures 4** and **5**, respectively.

Figure 4 shows that by increasing the dilution rate, the time required to establish a steady state decreased. At lower flow rates, $D = 0.1\ d^{-1}$ (**Figure 4a**), the time needed to reach a steady state was over 25 days, but when the dilution rate increased ($D = 0.8$ and $1.2\ d^{-1}$), time decreased by up to 4–6 days, which can be seen in **Figure 4b** and **c**. At the higher flow rates (**Figure 4d**), however, this time can be shorter than 2.5 days. Whereas the microorganisms at the beginning of the operation needed to adapt to a new environment, the amount of biomass was low and the substrate concentration was high, and consequently less substrate was converted. After a while, the value of the substrate was reduced (because of increased consumption) and the biomass increased to a value corresponding to a steady state. The biomass concentration in the steady state increased when increasing the flow but only up to a certain limit. **Figure 4b** shows that a steady state was achieved after approximately 6 days of continuous operating. The

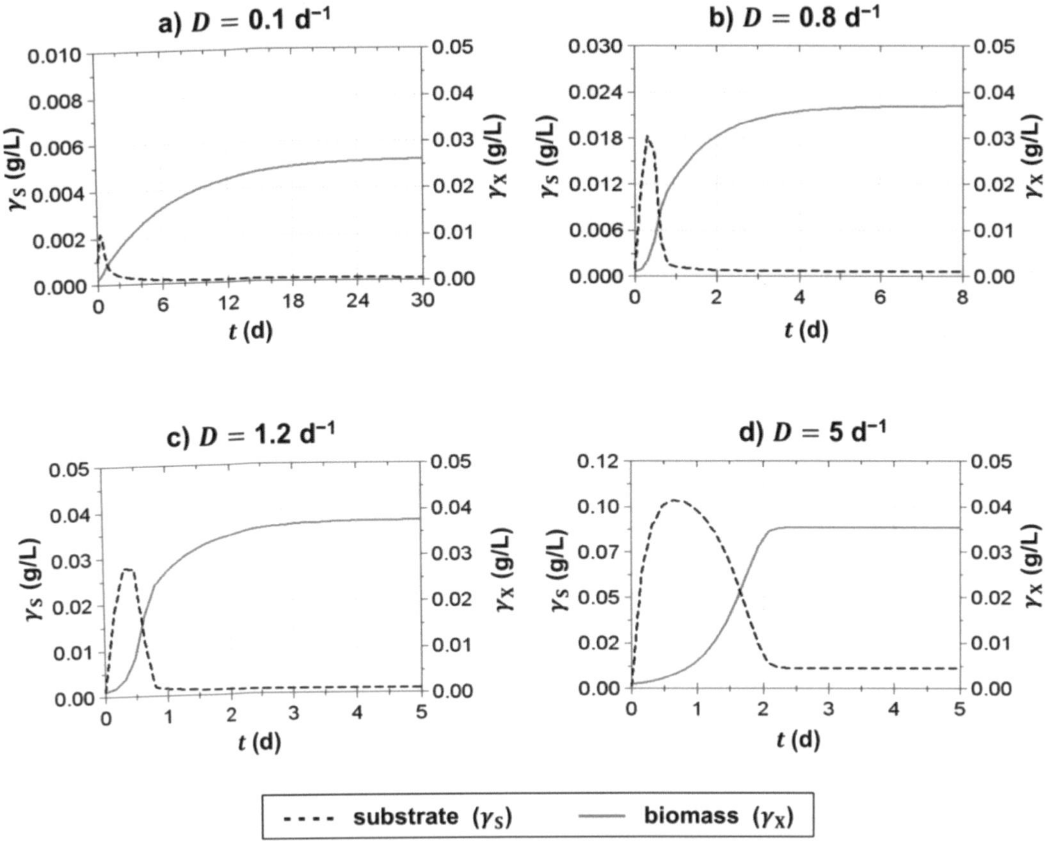

Figure 4. Dynamic concentration profiles for substrate ($\gamma_S = 0.001$ g/L) and active biomass at four different dilution rates: (a) $D = 0.1\ d^{-1}$, (b) $D = 0.8\ d^{-1}$, (c) $D = 1.2\ d^{-1}$ and (d) $D = 5\ d^{-1}$.

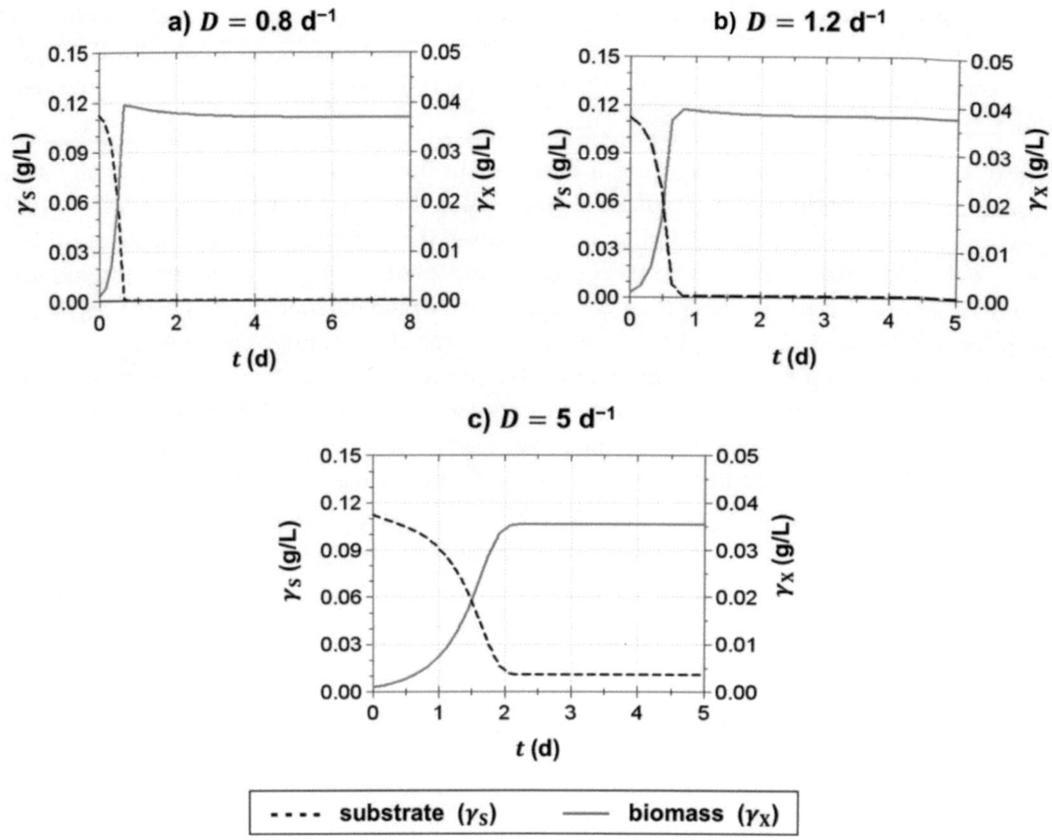

Figure 5. Dynamic concentration profiles (γ_S = 0.1126 g/L) for substrate and active biomass at three different dilution rates: (a) $D = 0.8\ d^{-1}$, (b) $D = 1.2\ d^{-1}$ and (c) $D = 5\ d^{-1}$.

active biomass and substrate concentrations in the steady state were 37 and 0.8 mg/L, respectively. At a dilution rate of 1.2 d^{-1} (**Figure 4c**), a steady state was achieved in 4 days. The concentration of substrate in the steady state at this dilution rate increased to 1.5 mg/L, whilst the concentration of biomass was quite similar. At higher dilution rates, a steady state was achieved even faster, but the substrate concentration in the steady state increased up to 9 mg/L and the biomass concentration decreased up to 35 mg/L.

During the final phase of our research, we developed the second model based on actual outlet substrate concentration, 0.1126 g/L. By comparing **Figures 4** and **5**, it can be seen that the outlet substrate concentration at the start of an operation has an insignificant impact on the concentration of active biomass and substrate in the steady state. It caused a change in the shape of the profile only at the beginning of the operation. The times required to reach steady states (for each dilution rate) were practically the same as presented in **Figure 4**. **Figure 5a** shows that a steady state was achieved in approximately 5–6 days, which is almost the same as presented in **Figure 4b**. The same applied for dilution rate 5 d^{-1}, where the times needed to reach a steady state in both cases were shorter than 2.5 days (**Figures 4d** and **5c**). Therefore, it can be concluded that outlet substrate concentration at the start of an operation has an insignificant impact on the final concentration of active biomass and substrate in the steady state.

4. Conclusion

Groundwater denitrification using a Zenon ZW 10 membrane bioreactor was studied and the validity verified regarding Monod kinetics for microbial growth. The research was carried out in two parts: firstly, the Monod kinetic parameters were determined by numerical interpolation of the experimental results and secondly, dynamic simulation was performed. The kinetic parameters obtained were 0.31 h^{-1} for the maximum specific growth rate of the biomass and 5.4 mg/L for the half-saturation constant. The calculated value of the yield coefficient was determined to be 35%. Using dynamic concentration profiles, the impact of the dilution rate on the substrate and biomass concentration was followed and the time required to establish a steady state was estimated. The results of dynamic simulation show that increase of the dilution rate decreased the time required to reach a steady state and that outlet substrate concentration has no significant impact on the concentration of the active biomass and substrate in the steady state.

Nomenclature

D	Dilution rate (h^{-1})
K_S	Half-saturation constant as COD (g/L)
q_D	Volume flow rate at the inflow (and at the outflow in the case of substrate mass balance) (L/h)
q_X	Volume flow rate at the outflow (L/h)
q_R	Volume flow rate at the recycle (L/h)
V	Reactor volume (L)
γ_D	Inflow mass concentration of biomass (g/L)
γ_X	Mass concentration of biomass in the reactor and at the outflow (g/L)
γ_R	Mass concentration of biomass at the recycle (g/L)
$\gamma_{S,D}$	Inflow mass concentration of substrate (g/L)
γ_S	Mass concentration of substrate at the outflow (g/L)
μ	Specific growth rate of biomass (h^{-1})
μ_{max}	Maximum specific growth rate of biomass (h^{-1})
$Y_{X/S}$	Yield coefficient (biomass regarding substrate) (g/g)
w_X	Percentage of active biomass (%)
k_d	Endogenous decay rate (h^{-1})

COD	Chemical oxygen demand (mg/L)
TSS	Total suspended solids (g/L)
MLSS	Mixed liquor suspended solids (g/L)

Author details

Marjana Simonič*, Andreja Goršek and Aleksandra Petrovič

*Address all correspondence to: marjana.simonic@um.si

Faculty of Chemistry and Chemical Engineering, University of Maribor, Maribor, Slovenia

References

[1] An S, Stone H, Nemati M. Biological removal of nitrate by an oil reservoir culture capable of autotrophic and heterotrophic activities: Kinetic evaluation and modeling of heterotrophic process. Journal of Hazardous Materials. 2011;**190**:686-693

[2] Gómez MA, González-López J, Hontoria-García E. Influence of carbon source on nitrate removal of contaminated groundwater in a denitrifying submerged filter. Journal of Hazardous Materials. 2000;**80**:69-80

[3] Karanasios KA, Vasiliadou IA, Pavlou S, Vayenas DV. Hydrogenotrophic denitrification of potable water: A review. Journal of Hazardous Materials. 2010;**180**:20-37

[4] Fernández-Nava Y, Marañón E, Soons J, Castrillón L. Denitrification of high nitrate concentration wastewater using alternative carbon sources. Journal of Hazardous Materials. 2010;**173**:682-688

[5] Wąsik E, Bohdziewicz J, Błaszczyk M. Removal of nitrates from groundwater by a hybrid process of biological denitrification and microfiltration membrane. Process Biochemistry. 2001;**37**:57-64

[6] Lee NM, Welander T. The effect of different carbon sources on respiratory denitrification in biological wastewater treatment. Journal of Fermentation and Bioengineering. 1996;**82**: 277-285

[7] Shen J, He R, Han W, Sun X, Li J, Wang L. Biological denitrification of high-nitrate wastewater in a modified anoxic/oxic-membrane bioreactor (A/O-MBR). Journal of Hazardous Materials. 2009;**172**:595-600

[8] Sears KJ, Alleman JE, Gong WL. Feasibility of using ultrasonic irradiation to recover active biomass from waste activated sludge. Journal of Biotechnology. 2005;**119**:389-399

[9] Her J-J, Huang J-S. Influences of carbon source and C/N ratio on nitrate/nitrite denitrification and carbon breakthrough. Bioresource Technology. 1995;**54**:45-51

[10] Glass C, Silverstein J. Denitrification kinetics of high nitrate concentration water: pH effect on inhibition and nitrite accumulation. Water Research. 1998;**32**:831-839

[11] Ravindran V, Tsai H-H, Williams MD, Pirbazari M. Hybrid membrane bioreactor technology for small water treatment utilities: Process evaluation and primordial considerations. Journal of Membrane Science. 2009;**344**:39-54

[12] Vacková L, Srb M, Stloukal R, Wanner J. Comparison of denitrification at low temperature using encapsulated Paracoccus denitrificans, Pseudomonas fluorescens and mixed culture. Bioresource Technology. 2011;**102**:4661-4666

[13] Oh J, Silverstein J. Oxygen inhibition of activated sludge denitrification. Water Research. 1999;**33**:1925-1937

[14] Sözen S, Çokgör EU, Orhon D, Henze M. Respirometric analysis of activated sludge behaviour—II. Heterotrophic growth under aerobic and anoxic conditions. Water Research. 1998;**32**:476-488

[15] Nuhoglu A, Pekdemir T, Yildiz E, Keskinler B, Akay G. Drinking water denitrification by a membrane bio-reactor. Water Research. 2002;**36**:1155-1166

[16] McAdam EJ, Judd SJ. A review of membrane bioreactor potential for nitrate removal from drinking water. Desalination. 2006;**196**:135-148

[17] Fenu A, Guglielmi G, Jimenez J, Spèrandio M, Saroj D, Lesjean B, Brepols C, Thoeye C, Nopens I. Activated sludge model (ASM) based modelling of membrane bioreactor (MBR) processes: A critical review with special regard to MBR specificities. Water Research. 2010;**44**:4272-4294

[18] Soto O, Aspé E, Roeckel M. Kinetics of cross-inhibited denitrification of a high load wastewater. Enzyme and Microbial Technology. 2007;**40**:1627-1634

[19] Vasiliadou IA, Pavlou S, Vayenas DV. A kinetic study of hydrogenotrophic denitrification. Process Biochemistry. 2006;**41**:1401-1408

[20] Akunna JC, Bizeau C, Moletta R. Nitrate and nitrite reductions with anaerobic sludge using various carbon sources: Glucose, glycerol, acetic acid, lactic acid and methanol. Water Research. 1993;**27**:1303-1312

[21] McAdam EJ, Judd SJ. Denitrification from drinking water using a membrane bioreactor: Chemical and biochemical feasibility. Water Research. 2007;**41**:4242-4250

[22] Sison NF, Hanaki K, Matsuo T. High loading denitrification by biological activated carbon process. Water Research. 1995;**29**:2776-2779

[23] Sison NF, Hanaki K, Matsuo T. Denitrification with external carbon source utilizing adsorption and desorption capability of activated carbon. Water Research. 1996;**30**:217-227

[24] Lee K-C, Rittmann BE. Applying a novel autohydrogenotrophic hollow-fiber membrane biofilm reactor for denitrification of drinking water. Water Research. 2002;**36**:2040-2052

[25] Zhao Y, Zhang B, Feng C, Huang F, Zhang P, Zhang Z, Yang Y, Sugiura N. Behavior of autotrophic denitrification and heterotrophic denitrification in an intensified biofilm-electrode reactor for nitrate-contaminated drinking water treatment. Bioresource Technology. 2012;**107**:159-165

[26] Buttiglieri G, Malpei F, Daverio E, Melchiori M, Nieman H, Ligthart J. Denitrification of drinking water sources by advanced biological treatment using a membrane bioreactor. Desalination. 2005;**178**:211-218

[27] Aslan Ş. Combined removal of pesticides and nitrates in drinking waters using biodenitrification and sand filter system. Process Biochemistry. 2005;**40**:417-424

[28] Li X, Chu HP. Membrane bioreactor for the drinking water treatment of polluted surface water supplies. Water Research. 2003;**37**:4781-4791

[29] McAdam E, Judd S. Optimisation of dead-end filtration conditions for an immersed anoxic membrane bioreactor. Journal of Membrane Science. 2008;**325**:940-946

[30] Tian J, Liang H, Nan J, Yang Y, You S, Li G. Submerged membrane bioreactor (sMBR) for the treatment of contaminated raw water. Chemical Engineering Journal. 2009;**148**:296-305

[31] Sanaeepur H, Hosseinkhani O, Kargari A, Ebadi Amooghin A, Raisi A. Mathematical modeling of a time-dependent extractive membrane bioreactor for denitrification of drinking water. Desalination. 2012;**289**:58-65

[32] Ergas SJ, Rheinheimer DE. Drinking water denitrification using a membrane bioreactor. Water Research. 2004;**38**:3225-3232

[33] Nalcaci OO, Böke N, Ovez B. Potential of the bacterial strain Acidovorax avenae subsp. avenae LMG 17238 and macro algae Gracilaria verrucosa for denitrification. Desalination. 2011;**274**:44-53

[34] Al-Malack MH. Determination of biokinetic coefficients of an immersed membrane bioreactor. Journal of Membrane Science. 2006;**271**:47-58

[35] Doran PM. Bioprocess Engineering Principles. Elsevier Science & Technology Books. 2004

[36] Shuler ML, Kargi F. Bioprocess Engineering, Basic Concepts. 2nd ed. Prentice Hall . 2002

[37] Peng J, Xue G. Mathematical modeling of hollow-fiber membrane system in biological wastewater treatment. Journal of Systemics, Cybernetics and Informatics. 2006;**4**:47-52

[38] Di Trapani D, Capodici M, Cosenza A, Di Bella G, Mannina G, Torregrossa M, Viviani G. Evaluation of biomass activity and wastewater characterization in a UCT-MBR pilot plant by means of respirometric techniques. Desalination. 2011;**269**:190-197

[39] Blanch WH, Clark SD. Biochemical Engineering. CRC press. 1997

[40] Bailey JE, Ollis DF. Biochemical Engineering Fundamentals. 2nd ed. McGraw-Hill. 1986

[41] Chung YC, Neethling JB. Viability of anaerobic digester sludge. Journal of Environmental Engineering ASCE. 1990;**116**:330-342

[42] Strotmann UJ, Geldem A, Kuhn A, Gendig C, Klein S. Evaluation of a respirometric test method to determine the heterotrophic yield coefficient of activated sludge bacteria. Chemosphere. 1999;**38**:3555-3570